AN INTRODUCTION TO ERROR CORRECTING CODES WITH APPLICATIONS

THE KLUWER INTERNATIONAL SERIES IN ENGINEERING AND COMPUTER SCIENCE

COMMUNICATIONS AND INFORMATION THEORY

Consulting Editor

Robert Gallager

Other books in the series:

Digital Communication. Edward A. Lee, David G. Messerschmitt.
ISBN 0-89838-274-2.

An Introduction to Cryptology. Henk C. A. van Tilborg.
ISBN 0-89838-271-8.

Finite Fields for Computer Scientists and Engineers. Robert J. McEliece.
ISBN 0-89838-191-6.

AN INTRODUCTION TO ERROR CORRECTING CODES WITH APPLICATIONS

by

Scott A. Vanstone
University of Waterloo

and

Paul C. van Oorschot
Bell Northern Research

KLUWER ACADEMIC PUBLISHERS
Boston/Dordrecht/London

Distributors for North America:
Kluwer Academic Publishers
101 Philip Drive
Assinippi Park
Norwell, Massachusetts 02061 USA

Distributors for all other countries:
Kluwer Academic Publishers Group
Distribution Centre
Post Office Box 322
3300 AH Dordrecht, THE NETHERLANDS

Library of Congress Cataloging-in-Publication Data

Vanstone, Scott A.
An introduction to error correcting codes with applications / by Scott A. Vanstone and Paul C. van Oorschot.
p. cm. — (The Kluwer international series in engineering and computer science. Communications and information theory)
Bibliography: p.
Includes index.
ISBN 978-1-4419-5117-5
1. Error-correcting codes (Information theory) 2. Data transmission systems. I. Van Oorschot, Paul C. II. Title. III. Series.
TK5102.5.V32 1989
005.7'2—dc20 89-31032
CIP

Printed in the United States of America

This printing is a digital duplication of the original edition.

Table of Contents

† This section may be omitted without loss of continuity.

Symbols and Notation

$[n,M]$-code	5	block code of block length n with M codewords
$\lfloor a \rfloor$	12	largest integer smaller than or equal to a
$\lvert X \rvert$	19	cardinality of set X
Z_n	22	integers modulo n
$a^{-1} \bmod n$	22	multiplicative inverse of a, modulo n; $a \cdot a^{-1} \equiv 1 (\bmod n)$
$Z_p[x]$	24	polynomials in indeterminate x, with coefficients from Z_p
$[g(x)]$	25,149	equivalence class containing polynomial $g(x)$
$Z_p[x]/(f(x))$	25,149	set of all equivalence classes in $Z_p[x]$ under congruence modulo the polynomial $f(x)$; sometimes denoted $Z_p[x]/f(x)$
$GF(q)$	28	Galois field (finite field) with q elements; q is a prime or a power of a prime number
$ord(\alpha)$	30	order of a field element
F^*	33	$F \backslash \{0\}$, the set of non-zero elements of F
$m_\alpha(x)$	33	minimal polynomial of α; sometimes $m_i(x)$ is used to denote $m_{\alpha^i}(x)$ or $m_{\beta^i}(x)$ (with α or β implied by the context)
$(a_0 a_1 a_2)$	35	vector representation of polynomial $a_0+a_1x+a_2x^2$
$z(i)$	39,277	Zech's logarithm of α^i; $\alpha^{z(i)}=1+\alpha^i$
α^∞	39	$\alpha^\infty=0$, by notational convention
$V_k(F)$	45	set of k-tuples over F
(n,k)-code	48	linear code of block length n and dimension k
(n,k,d)-code	48	linear code of block length n and dimension k, with distance d
$C^\perp$	54	orthogonal complement of linear code C ("C perp")
$\det D$	64	determinant of matrix D
$e^{2\pi i/p}$	89	$\cos(2\pi/p)+i \sin(2\pi/p)$, where $i^2=-1$
$R(1,r)$	115	first-order Reed-Muller code of block length 2^r
$R(k,r)$	144	kth-order Reed-Muller code of block length 2^r
$h_R(x)$	163	reciprocal polynomial of $h(x)$
C_i	181	cyclotomic coset (of q modulo n) containing i
$\mathbf{C}$	181	set of cyclotomic cosets (of q modulo n)
gcd	187,269	greatest common divisor

lcm	205	least common multiple
$\sigma(z)$	216	error locator polynomial for BCH decoding
$w(z)$	216	error evaluator polynomial for BCH decoding
$S(z)$	217	syndrome polynomial for BCH decoding
(n,k)–RS	237	(n,k) Reed-Solomon code
$\hat{\sigma}(z)$	244	error-erasure locator polynomial for BCH decoding
$\lambda(z)$	245	erasure locator polynomial for BCH decoding

Foreword

The field of Error Correcting Codes had its roots in Shannon's development of Information Theory in the 1940's. Research on coding was very active in the 50's and 60's and most of the major results known today were developed in that period. Most "practical engineers" of that era regarded coding as useless and impractical, while many of the researchers despaired of the field ever becoming more than an academic playground.

Applications for error correcting codes started in the 70's in very expensive communications systems such as deep space probes. As hardware costs dropped and as development engineers saw the benefits of real coding systems, applications expanded quickly. Today very sophisticated coding is used in ordinary consumer compact disc players, in storage technology, and in a broad range of communication systems.

This text develops the topic of error correcting block codes at an undergraduate level. It manages to do this both by being rather selective in the set of topics covered and by having an unusually clear organization and style of presentation. There are some sophisticated mathematical topics in the text and the authors develop them honestly, but the arguments are made accessible by using insight rather than excessive formalism.

One might wonder about the merits of an undergraduate course on error correcting block codes given the intense pressure to include more and more material in the curriculum for undergraduates in engineering, computer science, and mathematics. Wouldn't it be preferable to have a course covering convolutional codes as well as block coding and also covering many other communication topics? The type of course represented by this text, however, has many important advantages over a survey type course, not least of which is that survey courses are invariably boring. The major advantage of studying error correcting codes is the beauty of this particular combination of mathematics and engineering. The student, even if never involved with the field later, gets an example of applied mathematics and engineering at their best.

Robert G. Gallager, Consulting Editor
Fujitsu Professor of Electrical Engineering
Massachusetts Institute of Technology

Preface

An Introduction to Error Correcting Codes with Applications is a text intended for an introductory undergraduate course in error-correcting codes. It should be of interest to a wide range of students, from mathematics majors to those in computer science and engineering. The book deals almost entirely with codes that are linear in nature, and the necessary algebraic tools are developed, where necessary, to make the presentation self-contained. It is intended for students who are familiar with the fundamental concepts of linear algebra (solution of systems of linear equations, vector spaces, etc.). Background in abstract algebra (groups, rings and fields) would be of use, but is not assumed. In general, interested students with reasonable mathematical maturity will find the material to be at a suitable level.

While the underlying mathematics is completely developed, the book emphasizes the practical considerations of efficient encoding and decoding schemes for various codes. Most books in the area either emphasize the theoretical aspects of the subject (to the exclusion of practical considerations), or go to the other extreme, presenting implementation details at the circuit level. It is the purpose of this book to fill the void between the two, considering codes of practical interest, but stopping short of hardware-level implementation details. Codes which cannot be efficiently implemented are given little emphasis, and the temptation to discuss non-practical codes with "cute" mathematical properties has been resisted.

The book covers the basic principles involved in error-coding, and presents a thorough introduction to linear codes. It introduces cyclic codes in general, and progresses to the more sophisticated BCH codes, and the very practical Reed-Solomon codes. Applications of error-correcting codes to deep-space probe communications and to the increasingly popular *compact disc* players are emphasized. The goal is to provide a basic understanding of practical error-correcting codes and the principles they rely on. The development of an appreciation for the elegant mathematical concepts making efficient implementations possible is a most welcome byproduct.

The majority of the material presented is widely available in standard references, in various forms. Many of these we have included in our list of references; a most extensive bibliography can be found in [MacWilliams 77], although this of course does not contain the more recent results. Some results which we have included are less widely available, and are perhaps worth singling out. These include the results by van Lint and Wilson on bounds for cyclic codes (§6.3); the material dealing with coding techniques used in commercial compact disc players (parts of Chapter 7); and some factoring techniques (§5.9, §6.5) due to Berlekamp, which can be found in [Berlekamp 68] but are not typically included in other coding theory books.

Because of the importance of finite fields to a serious introduction to algebraic codes, the text provides a basic but thorough introduction to finite fields, assuming no prior knowledge of field theory. With the exception of a few sections which may be omitted at the instructor's discretion, the material can be covered in a one-semester course (thirteen weeks of lectures). It has been taught at the University of Waterloo for the past eight years to junior and senior undergraduate mathematics, computer science and engineering students. To aid instructors, we have denoted with a dagger (†) those sections which we feel may be omitted safely, if time is short. Inclusion of these sections, together with some of the theory developed in the exercise sets (eg. further bounds for linear codes, higher-order Reed-Muller codes, shortened Reed-Solomon codes), makes the text suitable for an introductory course at the graduate level, or for advanced undergraduates.

We have made a conscious effort to limit the amount of material in this book to that which one can reasonably expect to grasp in an introductory course. Unfortunately, this has meant that some areas meriting discussion have been left out. Perhaps the most notable of these is an introduction to convolutional codes, for which we refer the reader to [Blahut 83] or [Lin 83].

The text contains over 300 exercises, which range from routine to very challenging. We feel this is an essential component of an introductory course. A separate solution manual giving complete and detailed solutions to all exercises is currently being written.

Acknowledgements

The authors would like to thank the many people who took the time to constructively criticize this book in its various stages of development, point out simpler proofs to several standard results, and provide us with further sources of information. Among these people are Robert Day, Dieter Jungnickel, Therese Lee, Stan Lipshitz, Peter Manson, Bill McCuaig, Haluk Oral, Alfred Menezes, Ken Rosen, Doug Stinson, and Dan Wevrick.

We would also like to thank the editorial board of Kluwer Academic Publishers for many helpful suggestions. In particular, we gratefully acknowledge comments by R. Gallagher and R. McEliece. We would be remiss not to acknowledge the encouragement and support of this project by R. Holland, the publisher.

Finally, without the skillful typesetting of Christy Gillin, Joan Selwood and Marg Ward, the book would not exist. We thank them for all of their help and give special thanks to Joan Selwood for a great deal of extra effort during the final hectic days of preparation.

Chapter 1

INTRODUCTION and FUNDAMENTALS

1.1 An Introduction to Coding Theory

The *theory of error detecting and correcting codes* is that branch of engineering and mathematics which deals with the reliable transmission and storage of data. Information media are not 100% reliable in practice, in the sense that *noise* (any form of interference) frequently causes data to be distorted. To deal with this undesirable but inevitable situation, some form of *redundancy* is incorporated in the original data. With this redundancy, even if errors are introduced (up to some tolerance level), the original information can be recovered, or at least the presence of errors can be detected. A small example serves to illustrate the concepts.

Suppose the information we are to transmit comes from the set of symbols $\{A,B,C,D\}$. For practical considerations we associate sequences of 0's and 1's with each of these symbols.

$$A \to 00$$
$$B \to 10$$
$$C \to 01$$
$$D \to 11$$

Hence, if we send a 00 sequence, the receiver is to interpret this as the message A. The process can be represented by the following simple schematic diagram.

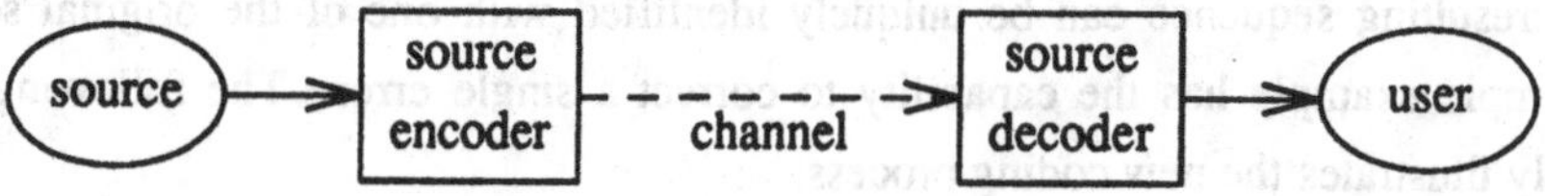

The source emits symbols from the information set which in our example is $\{A,B,C,D\}$. The *source encoder* associates each piece of information with a binary (0,1) sequence and then transmits it. The *channel* may be any information medium. For example, it may be a radio wave channel, a microwave channel, a cable, a digital integrated circuit or a storage disk. The *source decoder* receives the binary sequences from the channel, converts them back to the source alphabet and then passes this data to the user.

If the channel were 100% reliable, that is, if whatever the source encoder put on the channel was precisely what the source decoder received, then there would be no need for *error correction*. Unfortunately, most channels are not entirely reliable and there is a certain probability that what the decoder receives is not what was sent. If the source encoder sends 00 and the source decoder receives 01, then a single error has occurred. The decoder has no way of knowing this since 01 is a valid piece of information. The decoder has no choice but to pass the symbol C to the user. The challenge is to improve the reliability of message transmission. We do this by adding redundancy to each message. A major problem in coding theory is to determine how to add this redundancy to messages in order to detect and possibly correct channel errors. In our example, we might make the following associations.

$$A \rightarrow 00 \rightarrow 00000$$
$$B \rightarrow 10 \rightarrow 10110$$
$$C \rightarrow 01 \rightarrow 01011$$
$$D \rightarrow 11 \rightarrow 11101$$

Now, if the receiver reads 01000 it knows that an error has occurred, since this sequence is not one which the encoder put on the channel. If we assume that errors are introduced randomly to the binary digits (*bits*) of the sequence, then it seems reasonable for the decoder to assume that the transmitted sequence was 00000, since the received sequence can be obtained from this by the introduction of a single error. At least two errors would have to occur to transform any one of the other three sequences to 01000. One can easily check that altering a single bit in any one of the above 5-bit sequences results in a unique sequence. Hence if (at most) a single bit is altered, the resulting sequence can be uniquely identified with one of the original sequences. Thus our simple example has the capability to correct a single error. The following diagram schematically illustrates the new coding process.

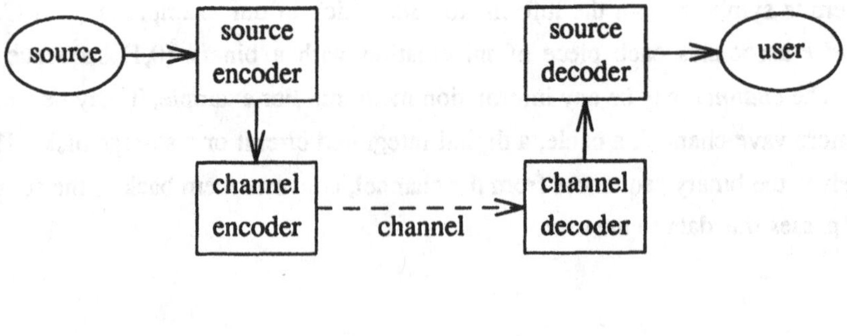

One of the problems to be resolved then is to determine how the channel encoder should add redundancy to the source encoder output. Another problem is to determine how the channel decoder should decide which sequence to decode to. We shall address these problems and others in later chapters. In §1.2, we indicate some specific applications of error-correcting codes, and in §1.3 we formalize the concepts just introduced.

The study of error-correcting codes is one branch of *coding theory*, a more general field of science dealing with the representation of data, including *data compression* and *cryptography*. These three areas are related in that they involve the transformation of data from one representation to an alternate representation and back again via appropriate *encoding* and *decoding* rules.

The objective of data compression is to transform data into a representation which is more compact yet maintains the information content (meaning) of the original data, to allow for more *efficient* use of storage and transmission media. This is possible by removing redundancy from the data. The traditional goal of cryptography has been to ensure *privacy* in communication by transforming data to render it unintelligible to all but the intended recipient. This is achieved through the use of an encoding scheme that relies on a *secret key* known only by the sender and intended recipient. As discussed above, the purpose of error-correcting codes is to increase the *reliability* of communication despite noise in the data medium, by adding redundancy in a uniform and efficient manner.

It is interesting to note that whereas cryptography strives to render data unintelligible to all but the intended recipient, error-correcting codes attempt to ensure data is decodable despite any disruption introduced by the medium. Data compression and error correction also contrast one another in that the former involves compaction and the latter data expansion. In this book we will be concerned mainly with error detecting and correcting codes.

If data compression and cryptographic encoding were to be used in conjunction with error-correcting codes, then the coding process as outlined in the above diagram would be modified slightly. A data compression stage might precede or replace the source encoding stage, and cryptographic encoding would be best performed just prior to channel encoding. Corresponding decoding stages would then be necessary at the appropriate points at the decoding end. Note that the removal of redundancy by data compression and subsequent addition of redundancy during channel encoding are not contradictory, as the redundancy that is removed by data compression serves no useful purpose, whereas that added by the channel encoder is of use for error control.

1.2 Applications of Error Correcting Codes

The increasing reliance on digital communication and the emergence of the digital computer as an essential tool in a technological society have placed error-correcting codes in a most prominent position. We cite here a few specific applications, in an attempt to indicate their practicality and importance.

The use of a *parity-bit* as an error-detecting mechanism is one of the simplest and most well-known schemes used in association with computers and computer communication. Data is partitioned into blocks of n bits, and then to each block, an additional bit is appended as a 0 or a 1, so as to make the number of bits which are 1 in the block, including the appended bit, an even number (for *even parity*). If during transmission then, a single bit-error occurs within the block, the parity of the block (i.e. the number of 1's in it) becomes odd, and checking the parity thus allows for detection of single errors.

Many computers now have error-correcting capabilities built into their random access memories; it is less expensive to compensate for errors through the use of error-correcting codes than to build integrated circuits that are 100% reliable. The single error-correcting *Hamming codes,* and *linear codes* in general, are of use here. These will be considered in Chapter 3. Disk storage is another area of computing where error-coding is employed. Storage capacity has been greatly increased through the use of disks of higher and higher density. With this increase in density, error probability also increases, and therefore information is now stored on many disks using error-correcting codes.

In 1972, the *Mariner* space probe flew past Mars and transmitted pictures back to earth. The channel for such transmissions is space and the earth's atmosphere. Solar activity and atmospheric conditions can introduce errors into weak signals coming from the spacecraft. In order that most of the pictures sent could be correctly recovered here on earth, the following coding scheme was used. The source alphabet consisted of 64 shades of grey. The source encoder encoded each of these into binary 6-tuples and the channel encoder produced binary 32-tuples. The source decoder could correct up to 7 errors in any 32-tuple. We shall discuss this system in more detail when we consider *Reed-Muller codes* in Chapter 4.

In 1979, the *Voyager* probes began transmitting colour pictures of Jupiter. For colour pictures the source alphabet needed to be much larger and was chosen to have 4096 colour shades. The source encoder produced binary 12-tuples for each colour shade and the channel encoder produced 24-tuples. This code would correct up to 3 errors in any 24-tuple, and is the *Golay code* discussed in Chapter 4.

The increasing popularity of *digital audio* is due in part to the powerful error-correcting codes that the digitization process facilitates. Information is typically stored on a small aluminized disk as a series of microscopic pits and smooth areas, the pattern representing a sequence of 0's and 1's. A laser beam is used as the playback mechanism to retrieve stored data. Because the data is digital, error correction schemes can be easily incorporated into such a system. Given an error-coded digital recording, a digital audio system can on playback, correct errors introduced by fingerprints, scratches, or even imperfections originally present in the storage medium. The *compact disc* system pioneered by Philips Corporation of the Netherlands in cooperation with Sony Corporation of Japan, allowing playback of pre-recorded digital audio disks, is an excellent example of the application of error-correcting codes to digital communication; *cross-interleaved Reed-Solomon* codes are used for error correction in this system. These ideas are pursued in Chapter 7. *Digital audio tape* (DAT) systems have also been developed, allowing digital recording as well as playback.

1.3 Fundamental Concepts

In this book we shall be primarily interested in *block codes*. We begin with a few definitions, in order to develop a working vocabulary. Let A be an *alphabet* of q symbols. For example, $A = \{a,b,c,...,z\}$ is the standard lower case alphabet for the English language, and $A = \{0,1\}$ is the *binary alphabet* used in the example of §1.1.

Definition. A *block code* of length n containing M *codewords* over the alphabet A is a set of M n-tuples where each n-tuple takes its components from A. We refer to such a block code as an $[n,M]$-code over A.

In practice, we most frequently take A to be the binary alphabet. We often refer simply to an $[n,M]$-code, the alphabet A being understood. We reserve round brackets ( ) for special types of codes, the *linear codes*, to be introduced in Chapter 3. Given a code C of block length n over an alphabet A, those specific n-tuples over A which are in C are referred to as *codewords*. Note that while the channel encoder transmits codewords, the n-tuples received by the channel decoder may or may not be codewords, due to the possible occurrence of errors during transmission. We use the terms *vector* and *word* interchangeably for n-tuple.

In §1.1, we constructed a [5,4]-code over a binary alphabet. That is, we constructed a code with 4 codewords, each being a 5-tuple (block length 5), with each component of the 5-tuple being 0 or 1.

The code is the set of n-tuples produced by the channel encoder (as opposed to the source encoder). The source encoder transforms messages into k-tuples over the code alphabet A, and the channel encoder assigns to each of these information k-tuples a codeword of length n. Since the channel encoder is adding redundancy, we have $n > k$ and hence we have *message expansion*. While the added redundancy is desirable from the point of view of error control, it decreases the efficiency of the communication channel by reducing its effective capacity. The ratio k to n is a measure of the fraction of information in the channel which is non-redundant.

Definition. The *rate* of an $[n,M]$-code which encodes information k-tuples is

$$R = \frac{k}{n}.$$

The rate of the simple code given in §1.1 is 2/5. We will consider rate more closely once we have introduced various other concepts. The quantity $r = n-k$ is sometimes called the *redundancy* of the code.

A fundamental parameter associated with an $[n,M]$-code C is the *Hamming distance* for C. Before we can define the Hamming distance for a code, we must define the Hamming distance between two codewords.

Definition. The *Hamming distance* $d(\mathbf{x},\mathbf{y})$ *between two codewords* $\mathbf{x}$ and $\mathbf{y}$ is the number of coordinate positions in which they differ.

Example 1.

Over the alphabet $A = \{0,1\}$, the codewords $\mathbf{x}$ and $\mathbf{y}$

$\mathbf{x} = (10110)$

$\mathbf{y} = (11011)$

have Hamming distance $d(\mathbf{x},\mathbf{y}) = 3$.

Example 2.

The codewords $\mathbf{u}$ and $\mathbf{v}$ over the alphabet $A = \{0,1,2\}$, given by

$\mathbf{u} = (21002)$

$\mathbf{v} = (12001)$

have Hamming distance $d(\mathbf{u},\mathbf{v}) = 3$.

It is easy to show that Hamming distance is a *metric* (exercise 4). That is, for all n-tuples $\mathbf{x}$, $\mathbf{y}$ and $\mathbf{z}$ over an alphabet A, Hamming distance satisfies the following three properties.

(1) $d(\mathbf{x},\mathbf{y}) \geq 0$, with equality if and only if $\mathbf{x} = \mathbf{y}$

(2) $d(\mathbf{x},\mathbf{y}) = d(\mathbf{y},\mathbf{x})$

(3) $d(\mathbf{x},\mathbf{y}) + d(\mathbf{y},\mathbf{z}) \geq d(\mathbf{x},\mathbf{z})$

Property (3) is often referred to as the *triangle inequality*. We can now define the *Hamming distance of a code*. Note how it is related to the Hamming distance between codewords.

Definition. Let C be an $[n,M]$-code. The *Hamming distance d of the code C* is

$$d = \min \{d(\mathbf{x},\mathbf{y}): \ \mathbf{x},\mathbf{y} \in C, \mathbf{x} \neq \mathbf{y}\}.$$

In other words, the Hamming distance of a code is the minimum distance between two distinct codewords, over all pairs of codewords.

Example 3.

Consider $C = \{\mathbf{c}_0, \mathbf{c}_1, \mathbf{c}_2, \mathbf{c}_3\}$ where

$\mathbf{c}_0 = (00000)$

$\mathbf{c}_1 = (10110)$

$\mathbf{c}_2 = (01011)$

$\mathbf{c}_3 = (11101)$

This code has distance $d = 3$. It is the [5,4] code constructed earlier. The reader may find it instructive to compute the distance between each pair of codewords.

Example 4.

Consider $C = \{\mathbf{c}_1,\mathbf{c}_2,\mathbf{c}_3,\mathbf{c}_4,\mathbf{c}_5\}$, a [5,5]-code over the alphabet $A = \{a,b,c,d,e\}$ with

$$\mathbf{c}_1 = (abcde)$$
$$\mathbf{c}_2 = (acbed)$$
$$\mathbf{c}_3 = (eabdc)$$
$$\mathbf{c}_4 = (bceda)$$
$$\mathbf{c}_5 = (adebc)$$

For $i \neq j$, we have $d(\mathbf{c}_i,\mathbf{c}_j) = 4$; the reader is invited to verify this. Thus $d = 4$.

As seen from these two examples, if one wants to compute the Hamming distance for an $[n,M]$-code C, then it is necessary to check $\binom{M}{2}$ pairs of codewords in order to find the pair with minimum distance. Note that there may be many pairs which give this minimum. In Chapter 3, we will see this work is simplified when the codes in question are *linear codes*.

Suppose we have an $[n,M]$-code C with distance d. We need to adopt a strategy for the channel decoder (henceforth called the decoder). When the decoder receives an n-tuple $\mathbf{r}$ it must make some decision. This decision may be one of

(i) no errors have occurred; accept $\mathbf{r}$ as a codeword.

(ii) errors have occurred; correct $\mathbf{r}$ to a codeword $\mathbf{c}$.

(iii) errors have occurred; no correction is possible.

In general, the decoder will not always make the correct decision; for example, consider the possibility of an error pattern occurring which changes a transmitted codeword into another codeword. The goal is that the decoder take the course of action which has the greatest probability of being correct (see discussion below). We shall make the assumption that errors are introduced by the channel at random, and that the probability of an error in one coordinate is independent of errors in adjacent coordinates. The decoding strategy we shall adopt, called *nearest neighbour decoding*, can then be specified as follows.

If an n-tuple $\mathbf{r}$ is received, and there is a unique codeword $\mathbf{c} \in C$ such that $d(\mathbf{r},\mathbf{c})$ is a minimum, then correct $\mathbf{r}$ to the $\mathbf{c}$. If no such $\mathbf{c}$ exists, report that errors have been detected, but no correction is possible.

By nearest neighbour decoding, a received vector is decoded to the codeword "closest" to it, with respect to Hamming distance. Again, we note that this strategy may on occasion lead to an incorrect decoding. We also note that this strategy sometimes leads to a decoding procedure which fails to decode some received vectors. Such a scheme is called an *incomplete decoding scheme*. To modify this scheme to a *complete decoding scheme* in which received vectors are always decoded (correctly or otherwise), in the event that a received vector is not closest to a unique codeword the vector could be decoded to any one of the codewords that is at a minimum distance from it, say one selected at random.

We rationalize the nearest neighbour decoding policy through the following considerations. Suppose the code is over an alphabet of q symbols. Let the probability that an error occurs on symbol transmission be p; the probability that a symbol is correctly transmitted over the channel is then $1-p$. We assume that if an error does occur, then each of the $q-1$ symbols aside from the correct symbol is equally likely to be received, with probability $p/(q-1)$ each. This hypothetical channel is called the *q-ary symmetric channel.*

Now suppose the vector **r** of block length n is received. The nearest neighbour decoding policy can be justified by the decoding strategy known as *maximum likelihood decoding*. Under this strategy, of all possible codewords **c**, **r** is decoded to that codeword **c** which maximizes the probability $P(\mathbf{r},\mathbf{c})$ that **r** is received, given that **c** is sent. If $d(\mathbf{r},\mathbf{c})=d$, then

$$P(\mathbf{r},\mathbf{c}) = (1-p)^{n-d}\left(\frac{p}{q-1}\right)^{d}.$$

To see this, note that $n-d$ coordinate positions in **c** are not altered by the channel, and this occurs with probability $(1-p)^{n-d}$. In each of the remaining d coordinate positions, the symbol in **c** is altered to that in **r**, and the probability of this is $p/(q-1)$ in each position.

Suppose now that $\mathbf{c}_1$ and $\mathbf{c}_2$ are two codewords, that **r** is received, and that

$$d(\mathbf{r},\mathbf{c}_1)=d_1, \qquad d(\mathbf{r},\mathbf{c}_2)=d_2.$$

Without loss of generality, assume that $d_1 \le d_2$. Now consider the conditions under which

$$P(\mathbf{r},\mathbf{c}_1) > P(\mathbf{r},\mathbf{c}_2).$$

If this holds then

$$(1-p)^{n-d_1}\left(\frac{p}{q-1}\right)^{d_1} > (1-p)^{n-d_2}\left(\frac{p}{q-1}\right)^{d_2}$$

$$(1-p)^{d_2-d_1} > \left(\frac{p}{q-1}\right)^{d_2-d_1}$$

and

$$\left(\frac{p}{(1-p)(q-1)}\right)^{d_2-d_1} < 1 .$$

If $d_1 = d_2$, this is false, and in fact, $P(\mathbf{r},\mathbf{c}_1) = P(\mathbf{r},\mathbf{c}_2)$. Otherwise, $d_2 - d_1 \geq 1$ and the inequality is true if and only if

$$\frac{p}{(1-p)(q-1)} < 1, \quad \text{i.e.} \quad p < \frac{(q-1)}{q} .$$

Hence if the probability that the channel transmits a symbol incorrectly is p, where $p < (q-1)/q$, then the codeword maximizing the probability that $\mathbf{r}$ is received is the codeword at a minimum distance from $\mathbf{r}$. We leave the case $p \geq (q-1)/q$ as an easy exercise (exercise 20).

If we assume that $0 \leq p < (q-1)/q$ then, the nearest neighbour decoding policy appears reasonable. (See exercise 21 for a further discussion.) We discuss the reliability of this technique at the end of this section.

Example 5.

Consider the binary code

$$C = \{ (00000), (10110), (01011), (11101) \}$$

and suppose that the symbol error probability for the channel is $p = 0.1$. If $\mathbf{r} = (11111)$ is received, note

$$P(\mathbf{r}, (00000)) = (0.1)^5 = 0.00001$$

$$P(\mathbf{r}, (10110)) = (0.1)^2(0.9)^3 = 0.00729$$

$$P(\mathbf{r}, (01011)) = (0.1)^2(0.9)^3 = 0.00729$$

$$P(\mathbf{r}, (11101)) = (0.1)^1(0.9)^4 = 0.06561$$

Since the probability $P(\mathbf{r}, (11101))$ is largest, $\mathbf{r}$ is decoded to (11101).

It may be that we are interested only in detecting when errors have occurred, without trying to correct them. If errors occur infrequently, and the source has the capability to retransmit on request, then this may be desirable. In other situations in which neither correction nor retransmission is possible, the knowledge that errors have occurred may nonetheless be useful (e.g. in digital audio systems as discussed in Chapter 7, where interpolation or muting can be used to conceal errors). A decoder used for error detection functions by simply reporting the occurrence of errors whenever a received word **r** is not a codeword.

A code is said to *correct e errors* if a decoder using the above scheme is capable of correcting any pattern of e or fewer errors introduced by the channel. In this case, the decoder can correct any transmitted codeword which has been altered in e or fewer coordinate positions. The significance of the Hamming distance for a code becomes apparent with the following theorem.

Theorem 1.1. Let C be an $[n,M]$-code having distance $d = 2e+1$. Then C can correct e errors. If used for error detection only, C can detect $2e$ errors.

Proof.

Let $\mathbf{c}_i$, $1 \le i \le M$ be the codewords of C. Let S be the set of all n-tuples over the alphabet of C and define

$$S_{\mathbf{c}_i} = \{\mathbf{x} \in S:\ d(\mathbf{x},\mathbf{c}_i) \le e\}.$$

$S_{\mathbf{c}_i}$ is called the *sphere of radius e* about the codeword $\mathbf{c}_i$. It consists of all n-tuples within distance e of the codeword $\mathbf{c}_i$, which we think of as being at the *centre* of the sphere. We first show that for $\mathbf{c}_i \ne \mathbf{c}_j$ we must have $S_{\mathbf{c}_i} \cap S_{\mathbf{c}_j} = \varnothing$. Suppose (by way of contradiction) that $\mathbf{x} \in S_{\mathbf{c}_i} \cap S_{\mathbf{c}_j}$. Then $d(\mathbf{x},\mathbf{c}_i) \le e$ and $d(\mathbf{x},\mathbf{c}_j) \le e$. By the triangle inequality for the Hamming metric,

$$d(\mathbf{c}_i,\mathbf{x}) + d(\mathbf{x},\mathbf{c}_j) \ge d(\mathbf{c}_i,\mathbf{c}_j)$$

and hence

$$d(\mathbf{c}_i,\mathbf{c}_j) \le 2e.$$

But every pair of distinct codewords have distance at least $2e+1$. Thus we conclude $S_{\mathbf{c}_i} \cap S_{\mathbf{c}_j} = \varnothing$. Hence if codeword $\mathbf{c}_i$ is transmitted and $t \le e$ errors are introduced, the received word **r** is an n-tuple in the sphere $S_{\mathbf{c}_i}$, and thus $\mathbf{c}_i$ is the unique codeword closest to **r**. The decoder can always correct any error pattern of this type.

If we use the code only for error detection, then at least $2e+1$ errors must occur in a codeword to carry it into another codeword. If at least 1 and at most $2e$ errors are introduced, the received word will never be a codeword and error detection is always possible. □

The notion of a *sphere of radius e* about each codeword provides a useful geometric model of the process of error correction. This model is particularly appealing with respect to *perfect codes,* which we mention briefly in Chapter 3.

In general, the distance of a code may be even or odd. The case for even distance is argued in a similar manner (exercise 7). We summarize both cases by stating the following result. As notation, let $\lfloor a \rfloor$ denote the largest integer smaller than or equal to a. For example, $\lfloor 3/2 \rfloor = 1$ and $\lfloor 11/2 \rfloor = 5$.

Theorem 1.2. Let C be an $[n,M]$-code with distance d. Then C can correct $\lfloor (d-1)/2 \rfloor$ errors. If used for detection only, C can detect $d-1$ errors.

Let us examine the case $d = 2e+1$ a bit more carefully. Suppose we can correct e errors. If we form the spheres $S_{\mathbf{c}_i}$ of radius e about each codeword $\mathbf{c}_i$, $1 \le i \le M$, it does not necessarily follow that every n-tuple of S is in one of these spheres. In fact, this generally does not occur, and hence there are n-tuples which are not contained in any of these spheres. If a vector $\mathbf{u}$ of this type is received by the decoder, it knows that at least $e+1$ errors have occurred. In general, no correction is possible but at least $e+1$ errors are detected. If C has distance $2e+1$, not all patterns of $e+1$ errors are detectable if the code is being used simultaneously for error correction, since it is then possible that $e+1$ errors introduced into a codeword $\mathbf{c}_i$ produce a received word in some sphere $S_{\mathbf{c}_j}$, $j \ne i$. In this case, the decoder would correct to $\mathbf{c}_j$, and not only would the $(e+1)$-error pattern go undetected, but a bogus "correction" would be performed. Hence if $d = 2e+1$, the code can correct e errors in general, but is unable to simultaneously detect additional errors. If the distance of the code is even, the situation changes slightly.

Theorem 1.3. Let C be an $[n,M]$-code having distance $d = 2k$. Then C can correct $k-1$ errors and simultaneously detect k errors.

Proof.

By Theorem 1.2, C can correct up to

$$\left\lfloor \frac{d-1}{2} \right\rfloor = \left\lfloor \frac{2k-1}{2} \right\rfloor = \left\lfloor k - \frac{1}{2} \right\rfloor = k - 1$$

errors. Since the spheres about codewords have radius $k-1$ any pattern of k errors cannot take a codeword into a word contained in some sphere about another codeword. Otherwise, the codewords at the centres of these two spheres would have distance at most $k+k-1 = 2k-1$, which is impossible since $d = 2k$. Hence, a received word which is obtained from a codeword by introducing k errors cannot lie in any codeword sphere and the decoder can detect the occurrence of errors. The decoder cannot detect $k+1$ errors in general, since a vector at distance $k+1$ from a given codeword may be in the sphere of another codeword, and the decoder would (erroneously) correct such a vector to the codeword at the centre of the second sphere. □

We illustrate the preceding concepts in the following example.

Example 6.

Consider the code $C = \{c_1, c_2, c_3, c_4\}$ of example 3 again, where

$$c_1 = (00000) \quad c_2 = (10110) \quad c_3 = (01011) \quad c_4 = (11101)$$

C has distance $d=3$, and hence can correct 1 error. The set S of all possible words over the alphabet $\{0,1\}$ consists of all possible binary 5-tuples. Hence, $|S| = 32$. The spheres about each codeword have radius 1.

$$S_{c_1} = \{\ (00000), (10000), (01000), (00100), (00010), (00001)\ \}$$

$$S_{c_2} = \{\ (10110), (00110), (11110), (10010), (10100), (10111)\ \}$$

$$S_{c_3} = \{\ (01011), (11011), (00011), (01111), (01001), (01010)\ \}$$

$$S_{c_4} = \{\ (11101), (01101), (10101), (11001), (11111), (11100)\ \}$$

These spheres of radius 1 cover precisely 24 of the 32 5-tuples in S. Let S^* be the set of 5-tuples not in any sphere.

$$S^* = \{\ (11000), (01100), (10001), (00101), (01110), (00111), (10011), (11010)\ \}.$$

Suppose the decoder receives $r = (00011)$. The distance to each codeword is computed.

$$d(c_1,r) = 2, \quad d(c_2,r) = 3, \quad d(c_3,r) = 1, \quad d(c_4,r) = 4.$$

Since there is a unique minimum distance, r is decoded to c_3. Notice that r lies in the sphere S_{c_3}.

Suppose the decoder receives $r = (11000)$. Distances are computed as follows.

$$d(\mathbf{c}_1,\mathbf{r})=2, \quad d(\mathbf{c}_2,\mathbf{r})=3, \quad d(\mathbf{c}_3,\mathbf{r})=3, \quad d(\mathbf{c}_4,\mathbf{r})=2.$$

The minimum distance is not unique. Since $\mathbf{r}$ is not at distance at most 1 from some codeword we can conclude that $\mathbf{r}$ is in no sphere and that at least 2 errors have occurred. Note that $\mathbf{r}$ is in S^*.

Suppose the codeword $\mathbf{c}_1$ is sent and 2 errors are introduced so that the received word is (10100). The decoder would (erroneously) correct this received word to $\mathbf{c}_2$. Note that $2 > \lfloor (d-1)/2 \rfloor = 1$. This code cannot detect all patterns of 2 errors if it is being used simultaneously for error correction.

If we are simply using the code C to detect errors with no intention of correcting, then we would be able to detect up to 2 errors. On receiving the vector (10100), the decoder would report the occurrence of errors.

This brings up an important point in the design of a code. In practice the code should be designed appropriately depending on the expected rate of errors for the particular channel being employed. If it is known that the probability that the channel will introduce 2 errors into any transmitted codeword is extremely small, then it is likely not necessary to construct a code that will correct all 2-bit error patterns. A single error correcting code will likely suffice. Conversely, if double errors are frequent and a single error correcting code is being used, then decoding errors will be frequent.

We now return to consider the reliability of nearest neighbour decoding. To be more precise, consider an $[n,M]$-code C having distance d over an alphabet A where $|A|=q$. Suppose the channel introduces a symbol error with probability p, where $0 \le p \le (q-1)/q$. What is the probability that a codeword $\mathbf{c}$ sent over the channel is correctly decoded at the receiving end by the decoder? To determine a lower bound on this probability, note that $\mathbf{c}$ will certainly be correctly decoded if the decoder receives any vector in the sphere of radius $e = \lfloor (d-1)/2 \rfloor$ about $\mathbf{c}$ (i.e. in the sphere $S_{\mathbf{c}}$ of radius e). The probability of this is

$$\sum_{\mathbf{r} \in S_{\mathbf{c}}} P(\mathbf{r},\mathbf{c}) = \sum_{i=0}^{e} \binom{n}{i} p^i (1-p)^{n-i},$$

giving a lower bound on the probability that a transmitted codeword is correctly decoded. Note however that using nearest neighbour decoding, if the channel introduces too many errors, the decoder will decode incorrectly, i.e. decode to a codeword other than that sent.

Example 7.

Consider the probability that nearest neighbour decoding applied to the code of example 5 decodes a transmitted codeword correctly. The decoder will certainly decode correctly if the channel introduces at most one error to any codeword (recall $d = 3$ here). For a given codeword, and symbol error probability $p = 0.1$, this occurs with probability

$$(0.9)^5 + \binom{5}{1}(0.1)^1(0.9)^4 = 0.91854$$

Suppose now that the channel error rate changes so that $p = 0.4$. The probability then becomes

$$(0.6)^5 + \binom{5}{1}(0.4)^1(0.6)^4 = 0.33696$$

Note that nearest neighbour decoding becomes less reliable as the symbol error rate increases. This is as one would expect.

1.4 Some Questions

From this brief introduction, a number of questions arise. Given n, M and d, can we determine if an $[n,M]$ code with distance d exists? Assuming such a code does exist, how would one be constructed in practice? How should information k-tuples be associated with codeword n-tuples to facilitate efficient channel encoding? How should channel decoding be performed?

Let us examine the issue of channel decoding in greater detail. We have outlined a simple decoding procedure in the preceding section. It involved computing the distance from the received word to each codeword, requiring M comparisons. While this might be acceptable for small M, in practice we usually find M quite large. Suppose a code C is used with $M = 2^{50}$ codewords, which is not unrealistic. If we could carry out 1 million distance computations per second, it would take around 20 years to make a single correction. Furthermore, it would take $\binom{M}{2}$ computations to determine the distance of the code, if we compute this distance by considering all codeword pairs to determine the pair with minimum distance. Clearly, this is not tolerable and more efficient techniques are required. Wishing to retain the nearest neighbour decoding strategy, we seek more efficient techniques for its implementation.

The chapters which follow shall be concerned with constructing codes for various values of n, M and d, and the consideration of appropriate encoding and decoding techniques. Most of the best known codes are algebraic in nature. In Chapter 2, we present a brief introduction to *finite fields,* which prove to be most important to the study of algebraic codes. Chapter 3 introduces the general class of *linear codes,* and Chapter 4 investigates two linear codes of particular interest. Chapter 5 examines some general characteristics of the more structured class of *cyclic codes*, and Chapter 6 introduces *BCH codes*, an important subclass of cyclic codes. The BCH codes known as *Reed-Solomon codes* are considered briefly in Chapter 7.

1.5 Exercises

1. Determine the rate of the code used by (a) the *Mariner*, and (b) the *Voyager* space probe.

2. Compute the distance between the following pairs of codewords.
(a) (0142), (3132) (b) (*aabca*), (*cabcc*) (c) $(\alpha\gamma\beta\lambda\gamma\chi)$, $(\gamma\beta\alpha\lambda\beta\psi)$

3. Consider a code C with $\mathbf{x} = (1001001)$ and $\mathbf{y} = (1011100)$ among its codewords.
(a) What is the Hamming distance between these codewords?
(b) What can we say about the distance of any code C containing these codewords?

4. Prove that Hamming distance between n-tuples is a metric.

5. Let C be an $[n,M]$-code of distance d. Suppose the codeword $\mathbf{c}$ is transmitted, and at most $\lfloor (d-1)/2 \rfloor$ errors are introduced by the channel, resulting in the received vector $\mathbf{r}$. Prove that $d(\mathbf{c},\mathbf{r}) < d(\mathbf{c}',\mathbf{r})$ for all other codewords $\mathbf{c}' \in C$, $\mathbf{c}' \neq \mathbf{c}$, and hence nearest neighbour decoding will correctly decoded $\mathbf{r}$ to $\mathbf{c}$.

6. Consider the code $C = \{(11000), (01101), (10110), (00011)\}$.

   (a) Construct a table with rows and columns indexed by codewords in C, and entries specifying the distance between each codeword pair. Hence determine the distance of C.

   (b) Decode the following received vectors, using nearest neighbour decoding.
   (i) $\mathbf{r}_1 = (01111)$ (ii) $\mathbf{r}_2 = (10110)$ (iii) $\mathbf{r}_3 = (11011)$ (iv) $\mathbf{r}_4 = (10011)$

7. Let C be an $[n,M]$-code having distance $d = 2t$.
(a) Prove that C can correct any pattern of $t-1$ errors in a codeword.
(b) Prove that if used for error-detection only, C can detect $2t-1$ errors.

8. Let C be an $[n,M]$-code having distance d. For (i) $d = 2t+1$, and (ii) $d = 2t$, state

   (a) the maximum number of errors that C can correct in general

   (b) the maximum number of errors that C can detect in general, if used for error-detection only

   (c) the maximum number of errors that C can detect in general, if used simultaneously for error-correction

9. Consider the code C of example 5. Redo the example "by picture".

   (a) Draw the spheres of radius $e = 1$ about each codeword. That is, about each codeword draw a circle (with the codeword at its centre), and within the circle put all n-tuples that differ in at most 1 coordinate position from that codeword.

(b) Verify that no n-tuple is in more than one sphere.

(c) How many n-tuples are in each sphere? How many n-tuples are in no sphere?

(d) Decode the following received vectors, by locating them within a sphere if possible, and correcting to the codeword at the centre of that sphere (i.e. assume at most 1 error occurred).
(i) $\mathbf{r}_1 = (00011)$ (ii) $\mathbf{r}_2 = (10100)$ (iii) $\mathbf{r}_3 = (11000)$

(e) Why could we not correct $\mathbf{r}_3$?

10. Discuss the relation and tradeoffs between rate and redundancy in an error-correcting code.

11. Is it possible for an error-correcting code to add sufficient redundancy so that a received vector can be correctly decoded to the transmitted codeword regardless of the number of errors occurring during the transmission? Explain.

12. Define the *Hamming weight* of an n-tuple $\mathbf{x}$ over an alphabet A to be the number of non-zero coordinates in $\mathbf{x}$, and denote this $w(\mathbf{x})$. Suppose the coordinate alphabet is $A = Z_m$, the integers modulo m, and addition and subtraction of n-tuples is componentwise.
(a) Prove that $d(\mathbf{x},\mathbf{y}) = w(\mathbf{x}-\mathbf{y})$.
(b) Determine all values of m such that $d(\mathbf{x},\mathbf{y}) = w(\mathbf{x}+\mathbf{y})$.

13. For binary n-tuples $\mathbf{x}$ and $\mathbf{y}$, and Hamming weight as defined above, prove the following.

(a) $d(\mathbf{x},\mathbf{y}) = w(\mathbf{x}+\mathbf{y}) = w(\mathbf{x}) + w(\mathbf{y}) - 2s$, where s is the number of coordinate positions in which both $\mathbf{x}$ and $\mathbf{y}$ are 1.

(b) $w(\mathbf{x}+\mathbf{y})$ is even if and only if $\mathbf{x}$ and $\mathbf{y}$ have the same parity (i.e. both have even weight or both have odd weight).

14. (a) Construct a binary [8,8]-code with distance 4.
(b) Construct a binary [8,16]-code with distance 4, or prove that no such code exists.

15. Construct a [6,9]-code over the alphabet $A = \{a,b,c\}$ with distance 4, or prove that no such code exists.

16. (a) Construct a binary [8,4]-code with distance 5.
(b) Construct a binary [7,3]-code with distance 5, or prove no such code exists.

17. Prove that a binary $[n,M]$-code C of distance $d = 2t+1$ exists if and only if a binary $[n+1,M]$-code C' of distance $d' = 2t+2$ exists.

18. Prove or disprove the following statement. A t-error-correcting code is incapable of correcting any pattern of $t+1$ errors.

19. Let C be a code with distance $d = 2t+u+1$. Determine the maximum number of errors that C can detect if used simultaneously to correct t errors.

20. In this question we further examine the maximum likelihood decoding strategy.

    (a) Prove that if the probability that the q-ary symmetric channel transmits a symbol incorrectly is $p = (q-1)/q$, then the probability that a vector $\mathbf{r}$ is received is the same regardless of the codeword $\mathbf{c}$ sent (and its distance from $\mathbf{r}$).

    (b) Prove that for $p > (q-1)/q$, the probability that the vector $\mathbf{r}$ is received is maximized by the codeword $\mathbf{c}$ at maximum distance from $\mathbf{r}$.

21. Suppose codewords are not necessarily sent with equal probability.

    (a) Show that nearest neighbour decoding does not necessarily maximize the probability of correct decoding, i.e. does not necessarily decode to the codeword $\mathbf{c}$ which maximizes the probability $P(\mathbf{c} \mid \mathbf{r})$ that $\mathbf{c}$ was sent, given that the vector $\mathbf{r}$ was received. An example will suffice.

    (b) The probability of correct decoding is maximized (the probability of incorrect decoding is minimized) by a strategy called *minimum error decoding*. By this strategy, a received vector $\mathbf{r}$ is decoded to the codeword $\mathbf{c}$ which maximizes the conditional probability $P(\mathbf{c} \mid \mathbf{r})$. Show that minimum error decoding and maximum likelihood decoding are equivalent (and both correspond to nearest neighbour decoding) if and only if all M codewords have equal *a priori* probability $1/M$.

22. Let $V_n(A)$ be the set of all n-tuples over an alphabet A. Let f be any bijection from A to $\mathbf{Z}_m$. Define a mapping ϕ: $V_n(A) \rightarrow V_n(\mathbf{Z}_m)$ by $\phi((x_1, x_2, ..., x_n)) = (f(x_1), f(x_2), ..., f(x_n))$. Prove that ϕ preserves Hamming distance, i.e. $d(\mathbf{x},\mathbf{y}) = d(\phi(\mathbf{x}),\phi(\mathbf{y}))$, for all $\mathbf{x},\mathbf{y} \in V_n(A)$. (Note: this result establishes that we can, without any loss of generality, think of the alphabet of m symbols as the integers modulo m whenever we are considering the Hamming metric on $V_n(A)$.)

23. Let $\mathbf{x},\mathbf{y} \in V_n(\mathbf{Z}_m)$ and let $X = \{\mathbf{u}\colon\ d(\mathbf{x},\mathbf{u}) < d(\mathbf{y},\mathbf{u})\}$ and $Y = \{\mathbf{u}\colon\ d(\mathbf{x},\mathbf{u}) > d(\mathbf{y},\mathbf{u})\}$. Prove that $|X| = |Y|$, where $|X|$ means the cardinality of the set X.

24. Suppose $C \subseteq V_n(\mathbf{Z}_m)$ with the property that if $\mathbf{x},\mathbf{y} \in C$ then $\mathbf{x}-\mathbf{y} \in C$. For each $\mathbf{x} \in C$, define the set $U_\mathbf{x} = \{\mathbf{u}\colon\ d(\mathbf{u},\mathbf{x}) < d(\mathbf{u},\mathbf{y})$ for all $\mathbf{y} \in C, \mathbf{y} \neq \mathbf{x}\}$. For $\mathbf{x},\mathbf{y} \in C$, $\mathbf{x} \neq \mathbf{y}$, prove that $|U_\mathbf{x}| = |U_\mathbf{y}|$.

25. Prove that the result given in exercise 24 does not necessarily hold if $|C| > 2$ and C is not closed under vector subtraction.

26. Let C be an $[n,M]$-code over $\mathbf{Z}_2$. Suppose that the decoder does nearest neighbour decoding with the additional rule that if the minimum distance is not unique then it will decode to a codeword by randomly selecting one of the codewords at minimum distance. Show by example that if we use this scheme with a channel symbol error rate $p = ½$ then decoding can in fact be worse than simply having the decoder randomly choose a codeword.

27. Consider the binary code

$$C = \{(0000101),(0011101),(1111100),(1111111),(0101011)\}.$$

Suppose that the symbol error probability for the channel is $p = .25$. If $\mathbf{r}_1 = (1101001)$ and $\mathbf{r}_2 = (0110101)$ are received, decode $\mathbf{r}_1$ and $\mathbf{r}_2$ by maximum likelihood decoding.

28. Consider an $[n-1,M]$-code C over an alphabet A having a zero element. Suppose we add an nth component, x_n, to each codeword $\mathbf{x} \in C$ such that

$$x_n = \begin{cases} 0 & \text{if } w(\mathbf{x}) \text{ is even} \\ 1 & \text{if } w(\mathbf{x}) \text{ is odd.} \end{cases}$$

Let the new code be C'. Suppose the channel has symbol error probability p. What is the probability that errors go undetected?

29. Let $C = \{\mathbf{c}_1,\mathbf{c}_2, \ldots, \mathbf{c}_{2^n}\}$ be an $[n,2^n]$ binary code such that the Hamming distance between c_i and c_{i+1}, $1 \le i \le 2^n-1$, is 1. (This is known as a *Gray code*.) Construct an $[n+1,2^{n+1}]$ binary code C' from C so that C' is also a Gray code.

30. Let $\mathbf{x}$ be a codeword of an $[n,M]$-code over an alphabet A with q symbols (A has a zero element). Suppose the weight of $\mathbf{x}$ is l (i.e. $w(\mathbf{x}) = l$). Determine the number of vectors of weight h which are at distance s from $\mathbf{x}$; this number is denoted $N(l,h;s)$. (This shows that the number of vectors of weight h at distance s from a given vector of weight l is independent of the given vector.)

Chapter 2

FINITE FIELDS

2.1 Introduction

In succeeding chapters we shall make extensive use of the algebraic structure known as a *finite field*. The reader already familiar with the elementary properties of finite fields may wish to proceed directly to Chapter 3. As the material regarding minimal polynomials (§2.4) and Zech's log tables (§2.6) is not referenced until §5.8, other readers may wish to bypass these sections until that time. For a more complete introduction to finite fields, the reader is invited to consult [McEliece 87], and for further details, the definitive reference [Lidl 84].

We begin here by giving a formal definition of a *field*.

Definition. A *field* F is a set of elements which is closed under two binary operations, which we denote by "+" and "·", such that the following axioms are satisfied for all $a, b, c \in F$.

(i) $a + (b+c) = (a+b) + c$

(ii) $a + b = b + a$

(iii) there exists an element $0 \in F$ such that $a + 0 = a$

(iv) there exists an element $-a \in F$ such that $a + (-a) = 0$

(v) $a \cdot (b \cdot c) = (a \cdot b) \cdot c$

(vi) $a \cdot b = b \cdot a$

(vii) there exists an element $1 \in F$ such that $a \cdot 1 = a$

(viii) for each $a \neq 0$, there exists an element $a^{-1} \in F$ such that $a \cdot a^{-1} = 1$

(ix) $a \cdot (b+c) = a \cdot b + a \cdot c$

Informally, a field is a set of elements closed under two operations we usually refer to as "addition" and "multiplication". Both operations are associative and commutative. Additive and multiplicative identities exist, elements have inverses with respect to both operations (with the exception that 0 has no multiplicative inverse), and the multiplication distributes over the addition "in the usual way".

More concisely, properties (i)-(iv) establish that the set is an *abelian group* under addition, with additive identity 0 (see §3.6). Properties (v)-(viii) establish that the non-zero elements form an abelian group under multiplication, with multiplicative identity 1. Property (ix) is the distributive property.

Examples of fields are the rational numbers, the real numbers, and the complex numbers. Each of these examples is a field with an infinite number of elements. If the field contains only a finite number of elements, it is called a *finite field*. It is easy to establish precisely when the integers modulo n, denoted Z_n, under the operations of standard addition and multiplication mod n, form a finite field.

Theorem 2.1. Z_n is a finite field if and only if n is a prime number.

Proof.

If n is not prime, then n has a nontrivial factorization $n = ab$. In this case, the element b has no multiplicative inverse, since $bc \equiv 1 \pmod{n}$ would imply $abc \equiv a \pmod{n}$, and then $0 \equiv a \pmod{n}$, contradicting the factorization of n. Hence if n is not prime, then Z_n is not a field.

Conversely, if n is prime, then all field axioms are easily verified, with the exception of the existence of multiplicative inverses. To establish this existence, one may employ the Euclidean algorithm for integers (see Appendix B). Given an element $0 < a < n$, since n is prime we have $gcd(n,a) = 1$; thus by the Euclidean algorithm, there exist integers s and t such that $s \cdot n + t \cdot a = 1$; that is, $t \cdot a \equiv 1 \pmod{n}$ and $t = a^{-1} \pmod{n}$. Establishing the existence of multiplicative inverses completes the proof that Z_n is a field if n is prime. Since Z_n has a finite number of elements, Z_n is a finite field in this case. □

For example, Z_2 and Z_5 are finite fields, but Z_9 is not, because 3 has no multiplicative inverse in Z_9. Much of the fundamental theory of error-correcting codes can be developed using only the finite fields Z_p (p a prime). However, we shall see shortly that fields besides these exist, and many of these others are used in practice. Hence, we proceed to consider finite fields in general.

Definition. Let F be a field. The *characteristic* of F is the least positive integer m such that

$$\sum_{i=1}^{m} 1 = 1 + 1 \cdots + 1 = 0,$$

where $1 \in F$ is the multiplicative identity of the field, and the summation represents addition in the field. If no such m exists, we define the characteristic to be 0.

For example, Z_2 has characteristic 2, and Z_5 has characteristic 5. It is easy to see that the field Z_p has characteristic p. It follows that in a field of characteristic p, any element added to itself p time yields 0, i.e. any element multiplied by p vanishes. For example, in $F=Z_5$, $\sum_{i=1}^{5} 1 = 5 \cdot 1 \equiv 0 \pmod 5$, and indeed $5 \cdot a \equiv 0 \pmod 5$ for all $a \in Z_5$. The rational, real and complex numbers are examples of fields of characteristic 0.

From this definition, it follows that the characteristic of a field cannot be a composite number.

Theorem 2.2. If the characteristic m of a field F is not 0, then m is a prime number.

Proof.

Suppose m is not a prime number. Then $m = ab$ where $a > 1$ and $b > 1$. Now, if we consider the field elements t and s, where

$$t = \sum_{i=1}^{a} 1 = 1 + 1 + \cdots + 1, \qquad s = \sum_{i=1}^{b} 1 = 1 + 1 + \cdots + 1$$

then $ts = \sum_{i=1}^{ab} 1 = \sum_{i=1}^{m} 1 = 0$ since the characteristic is m. Now $t \in F$ and $s \in F$, and neither is 0 (since $a, b < m$ and m is the characteristic), but $ts = 0$. Since $t \neq 0$, we know t^{-1} exists, and hence $t^{-1}ts = t^{-1}0 = 0$ implies $s=0$. This contradicts the minimality of m. We conclude that m is prime. □

If F is a finite field of characteristic p, then F contains a *subfield* (i.e. a subset of elements which is a field in its own right) having p elements. To see this, we consider the set of elements

$$1,\ 1+1,\ 1+1+1, \ldots, \sum_{i=1}^{p} 1.$$

Note that last element in the list is 0, and the first is 1. These elements are all distinct, closed under addition and multiplication, and the additive and multiplicative inverses can be shown to exist for the non-zero elements. This subfield is essentially the integers modulo p and so we will assume that any finite field of characteristic p contains Z_p as a subfield. Z_p is referred to as the *ground field* of F. A convenient way to think of F is as a *vector space* over the ground field Z_p. (For a review of vector spaces, see Appendix A.) That is, we think of the elements of F as *vectors* with the *scalars* being from Z_p. Under this interpretation the elements of Z_p are both vectors and scalars. Since F is finite the vector space must have finite dimension. If the dimension is n, then there exist n elements $\{\alpha_1,\alpha_2,...,\alpha_n\}$ of F which form a *basis* for F over Z_p, and hence

$$F = \{ \sum_{i=1}^{n} \lambda_i \alpha_i : \lambda_i \in Z_p \}.$$

This implies that F contains p^n elements for some positive integer n. We state this result as the following theorem.

Theorem 2.3. If F is a finite field of characteristic p, then F contains p^n elements for some positive integer n.

This tells us that any finite field contains a prime or prime power number of elements. Hence, for instance, there can be no finite field having 6 elements. We now show that for any prime number p and any positive integer n there exists a finite field having p^n elements.

For $n = 1$, the integers modulo p provide fields on p elements. For $n \geq 2$ we start by considering the set $Z_p[x]$ of all polynomials in x over the field Z_p, where each polynomial has finite degree. For example, for $p = 2$,

$$Z_2[x] = \{ 0, 1, x, 1+x, 1+x^2, x^2, x+x^2, 1+x+x^2, \cdots \}$$

is the set of all polynomials in x of finite degree, with coefficients from the field Z_2. Under polynomial addition and multiplication $Z_p[x]$ forms an algebraic system known as a *commutative ring with identity,* which is one property (viii) short of being a field. (We examine rings in greater detail in §5.2.) From $Z_p[x]$ select an *irreducible* polynomial $f(x)$ of degree n. Recall that a polynomial is irreducible in $Z_p[x]$ if it cannot be written as the product of two polynomials of $Z_p[x]$ each of positive degree. It is, of course, necessary to prove that it is always possible to select such a polynomial. For the time being we will assume that such a choice can be made (and pursue the matter further in §2.2). Then using $Z_p[x]$ and this selected irreducible polynomial, we will shortly be able to construct a field with p^n elements. First, a few more definitions are necessary. Let $F=Z_p$, and let $f(x)$, $g(x)$ and $h(x)$ be polynomials in $F[x]$.

Definition. $h(x)$ is said to be *congruent to $g(x)$ modulo $f(x)$* if and only if there exists a polynomial $l(x) \in F[x]$ such that

$$h(x) - g(x) = l(x)f(x).$$

We write $h(x) \equiv g(x) \pmod{f(x)}$.

The definition implies that $h(x)$ and $g(x)$ are congruent if $f(x)$ divides their difference. Equivalently, $h(x)$ and $g(x)$ leave the same remainder when divided by $f(x)$. This definition is valid for any polynomial $f(x)$ in $F[x]$, not just the irreducible ones. It is not hard to show that this congruence is an *equivalence relation* on $F[x]$ (exercise 13). An equivalence relation partitions the set that is defined on into subsets called *congruence classes*.

Definition. For a given $f(x) \in F[x]$, the *equivalence class containing $g(x) \in F[x]$* is

$$[g(x)] = \{\ h(x) \in F[x]:\ \ h(x) \equiv g(x) \pmod{f(x)}\ \}.$$

$[g(x)]$ is the set of all polynomials which leave the same remainder as $g(x)$ does when divided by $f(x)$. The use of an equivalence relation to define congruence classes is common, and we are perhaps more comfortable with the idea of congruence in the integer domain or among the integers modulo p.

We define addition and multiplication of congruence classes of polynomials as follows. For $g(x)$, $h(x) \in Z_p[x]$ define

$$[g(x)] + [h(x)] \ = \ [g(x)+h(x)]$$

and

$$[g(x)]\cdot[h(x)] = [g(x)\cdot h(x)].$$

We often omit the multiplication operator "$\cdot$", and simply use juxtaposition to indicate multiplication. If $Z_p[x]/f(x)$ is the set of all equivalences classes in $Z_p[x]$ under congruence modulo the irreducible polynomial $f(x)$, then it is not hard to show that the operations above defined on this set of classes form a finite field. Most of the field axioms follow immediately. Axiom (viii) is the only difficult one to verify. Since $[1]$ is the multiplicative identity we need to show that for any $[g(x)] \neq [0]$, there exists $[h(x)]$ such that $[g(x)][h(x)] = [1]$. This requires that $g(x)h(x) \equiv 1 \pmod{f(x)}$. Since $f(x)$ is irreducible and $f(x)$ does not divide $g(x)$, the greatest common divisor of $g(x)$ and $f(x)$ is 1. By the *Euclidean algorithm for polynomials* (see Appendix B) we know that there exist polynomials $s(x)$ and $t(x)$ such that

$$s(x)g(x) + t(x)f(x) = 1$$

or

$$s(x)g(x) \equiv 1 \pmod{f(x)}.$$

Hence

$$[s(x)][g(x)] = [1]$$

which establishes that every non-zero class has a multiplicative inverse.

We also claim that $Z_p[x]/f(x)$ contains p^n elements. Consider any class $[g(x)]$. By the *division algorithm for polynomials* (see Appendix B)

$$g(x) = q(x)f(x) + r(x) \quad \text{where} \quad r(x) = 0 \quad \text{or} \quad deg\ r(x) < deg f(x).$$

Hence $[g(x)] = [r(x)]$. $r(x)$ is called a *remainder polynomial*. Since $r(x) = 0$ or $deg\ r(x) < deg f(x)$, no two distinct remainder polynomials can be in the same class. Since

$$r(x) = \sum_{i=0}^{n-1} a_i x^i, \quad a_i \in Z_p$$

there are p choices for each a_i and so there are p^n distinct remainder classes. This shows that $Z_p[x]/f(x)$ contains p^n elements, and we can now construct fields with p^n elements for $n \geq 2$. Together with the knowledge that for $n = 1$ Z_p is a field with p^1 elements, this shows the existence of finite fields with p^n elements for every prime p and positive integer n.

Example 1.

We construct a field with 4 elements. We use $f(x) = x^2 + x + 1 \in Z_2[x]$, which is irreducible over Z_2. This polynomial defines the field

$$F = \{ [0], [1], [x], [1+x] \}.$$

Addition is ordinary polynomial addition over Z_2. For example, $[1+x] + [x] = [1+2x] = [1]$ since $2 \equiv 0$ in Z_2, the field having characteristic 2. Construction of the addition table is straightforward.

+	[0]	[1]	[x]	[$1+x$]
[0]	[0]	[1]	[x]	[$1+x$]
[1]	[1]	[0]	[$1+x$]	[x]
[x]	[x]	[$1+x$]	[0]	[1]
[$1+x$]	[$1+x$]	[x]	[1]	[0]

Multiplication is performed using ordinary polynomial multiplication in $Z_2[x]$, and then reducing the product by computing the remainder modulo $f(x)$. The reduction can be done by long division or using the relation $x^2 \equiv x+1$ to reduce the degree of the product below the degree of $f(x)$, which is 2 here. For example,

$$[1+x]\cdot[1+x] = [1+2x+x^2] = [1+x^2] = [1+(1+x)] = [x].$$

The multiplication table is listed below.

·	[0]	[1]	[x]	[$1+x$]
[0]	[0]	[0]	[0]	[0]
[1]	[0]	[1]	[x]	[$1+x$]
[x]	[0]	[x]	[$1+x$]	[1]
[$1+x$]	[0]	[$1+x$]	[1]	[x]

For the sake of convenience, we henceforth take the liberty of writing $g(x)$ in place of $[g(x)]$.

Example 2.

We construct a field with 9 elements. Take $f(x) = 1 + x^2 \in Z_3[x]$, which is irreducible over Z_3 as required. The field is

$$F = \{ 0, 1, 2, x, 2x, 1+x, 1+2x, 2+x, 2+2x \}.$$

Addition in the field is standard polynomial addition with coefficient arithmetic in Z_3. For example,

$$[2+x] + [2+2x] = [(2+x)+(2+2x)] = [4+3x] = [1].$$

The addition table is once again straightforward, and hence omitted.

For multiplication, we reduce the product polynomials modulo $f(x)$. Note here that $x^2+1 \equiv 0$ implies $x^2 \equiv -1 \equiv 2$, and we may replace x^2 by 2. For example,

$$[2x]\cdot[1+x] \;=\; [(2x)(1+x)] \;=\; [2x+2x^2] \;=\; [2x+4] \;=\; [2x+1].$$

The multiplication table follows.

$\cdot$	0	1	2	x	$2x$	$1+x$	$2+x$	$1+2x$	$2+2x$
0	0	0	0	0	0	0	0	0	0
1	0	1	2	x	$2x$	$1+x$	$2+x$	$1+2x$	$2+2x$
2	0	2	1	$2x$	x	$2+2x$	$1+2x$	$2+x$	$1+x$
x	0	x	$2x$	2	1	$x+2$	$2x+2$	$1+x$	$1+2x$
$2x$	0	$2x$	x	1	2	$1+2x$	$1+x$	$2+2x$	$2+x$
$1+x$	0	$1+x$	$2+2x$	$x+2$	$1+2x$	$2x$	1	2	x
$2+x$	0	$2+x$	$1+2x$	$2+2x$	$1+x$	1	x	$2x$	2
$1+2x$	0	$1+2x$	$2+x$	$1+x$	$2+2x$	2	$2x$	x	1
$2+2x$	0	$2+2x$	$1+x$	$1+2x$	$2+x$	x	2	1	$2x$

If F is a finite field with q elements, we usually denote it $GF(q)$, and refer to it as the *Galois field* with q elements. Note that q must have the form p^n, i.e. be a prime or a power of a prime. As noted earlier, a field $GF(p^n)$ is a *field of characteristic p*. The field Z_p is similarly denoted $GF(p)$.

2.2 Irreducible Polynomials

In order to construct a finite field having p^n elements using the method of the previous section, we require an irreducible polynomial of degree n in $GF(p)[x]$. The first question we might then ask is whether or not one always exists for each positive integer n. We begin to answer this question by considering the case $n = 2$. Let us try to prove that there always exists a quadratic irreducible polynomial in $GF(p)[x]$. In fact, let us look for a monic irreducible polynomial. By *monic* we mean that the non-zero coefficient of the highest power of x is 1. There are p^2 monic polynomials of degree 2 in $GF(p)[x]$. If one of these is reducible, it is the product of two degree-1 monic polynomials. There are precisely p monic degree-1 polynomials. Using these we can construct $\binom{p}{2}+p$ reducible monic quadratic polynomials. (We can choose any two distinct monic polynomials or the square of any particular monic polynomial.) Hence the number of monic irreducible quadratic polynomials is

$$I_2 = p^2 - \binom{p}{2} - p \;=\; \binom{p}{2} \;>\; 0, \quad p \geq 2.$$

This proves the existence of irreducible quadratics. In a similar but slightly more complicated fashion we can count the number of monic irreducible cubic polynomials, I_3. There are p^3 monic cubic polynomials in total. Reducible monic cubics can be constructed by multiplying either a monic polynomial with an irreducible quadratic, three distinct irreducible monics, a monic and the square of another distinct monic, or the cube of a monic. Therefore

$$I_3 = p^3 - p \cdot I_2 - \binom{p}{3} - 2\cdot\binom{p}{2} - p = \frac{p(p^2-1)}{3} > 0, \quad p \geq 2,$$

and hence at least one such polynomial exists. It is possible to derive a formula for the general situation and deduce that the number of monic irreducible polynomials of degree n is positive for each n, but this is considerably more difficult than the simple cases above, and is beyond the scope of this book (see exercise 7).

2.3 Properties of Finite Fields

In this section, we consider primitive elements in finite fields, and the order of elements. Let F^* denote the set of non-zero field elements, $F \setminus \{0\}$.

Definition. An element α in a finite field F is said to be a *generator* of F^*, or *primitive element* of F if

$$\{\alpha^i : i \geq 0\} = F^*.$$

That is, if α generates all non-zero field elements.

Example 3.

In example 2, $GF(9)$ was constructed using the irreducible $f(x) = x^2 + 1$ from $Z_3[x]$. To find a primitive element, we proceed by trial and error. Note that $\alpha = x$ is not a primitive element here. However, if we try $\alpha = 1 + x$, and use the multiplication table given, then

$$\begin{array}{ll} (1+x)^0 = 1 & (1+x)^4 = 2 \\ (1+x)^1 = 1+x & (1+x)^5 = 2+2x \\ (1+x)^2 = 2x & (1+x)^6 = x \\ (1+x)^3 = 1+2x & (1+x)^7 = 2+x. \end{array}$$

Hence, $\alpha = 1 + x$ is a generator for $GF(9)^*$.

We now show that every finite field contains a primitive element. First, we require a definition and a few preliminary results.

Lemma 2.4. For every non-zero element $\alpha \in GF(q)$, $\alpha^{q-1} = 1$. Furthermore, an element $\alpha \in GF(q^m)$ lies in $GF(q)$ itself if and only if $\alpha^q = \alpha$.

Proof.

Let the distinct non-zero elements of $GF(q)$ be $a_1, a_2, ..., a_{q-1}$. Consider αa_1, αa_2, ..., αa_{q-1}. These are all distinct, because otherwise for some $i \neq j$, $\alpha a_i = \alpha a_j$, which multiplied by α^{-1} yields $a_i = a_j$. Hence $\{\alpha a_1, \alpha a_2, ..., \alpha a_{q-1}\} = \{a_1, a_2, ..., a_{q-1}\}$. This implies

$$(\alpha a_1)(\alpha a_2) \cdots (\alpha a_{q-1}) = a_1 a_2 \cdots a_{q-1}$$

which yields

$$\alpha^{q-1} = 1.$$

This proves the first statement. Hence $\alpha \in GF(q)$ implies $\alpha^q = \alpha$, and every element of $GF(q)$ is a root of the polynomial $x^q - x$. To prove the second statement, we must show that $\alpha^q = \alpha$ implies $\alpha \in GF(q)$. It is easy to establish (exercise 17) that a polynomial of degree n over $GF(q)$ has at most n roots in $GF(q)$. Here, we have $n = q$, and the elements of $GF(q)$ provide q distinct roots of the polynomial $x^q - x$. We conclude that this polynomial has no other roots, and that $\alpha^q = \alpha$ implies $\alpha \in GF(q)$. □

Definition. The *order* of a non-zero element $\alpha \in GF(q)$, denoted $ord(\alpha)$, is the smallest positive integer t such that $\alpha^t = 1$.

Example 4.

Using $GF(9)$ as given in example 2, it is easy to verify that $\text{ord}(1+x) = 8$ and $\text{ord}(x) = 4$.

Lemma 2.4 shows that the order of every non-zero element is well defined. We now make a stronger statement.

Lemma 2.5. For every non-zero element $\alpha \in GF(q)$, ord(α) divides $q - 1$.

Proof.

Let $t = \text{ord}(\alpha)$. By the division algorithm for integers,

$$q - 1 = lt + r, \quad 0 \le r < t.$$

Thus

$$\alpha^{q-1} = \alpha^{lt+r} = \alpha^{lt}\alpha^{r}.$$

But $\alpha^{q-1} = 1$ and $\alpha^{t} = 1$ implying $\alpha^{r} = 1$. Since $t = \text{ord}(\alpha)$ we conclude that $r = 0$ and that t divides $(q-1)$. □

We are now in a position to prove that primitive elements always exist.

Theorem 2.6. In every field $F=GF(q)$ there exists a primitive element.

Proof.

Let λ be an element of $GF(q)$ which has largest order t. If $t = q-1$, then λ is a primitive element and we are done. Suppose $t < q-1$. Suppose also that the order of every other non-zero element of $GF(q)$ divides t. Then every non-zero element satisfies the polynomial equation

$$y^t - 1 = 0.$$

But this equation has at most t roots in a field and since $t < q-1$ we have a contradiction. Hence, there is some element $\beta \in GF(q)$ such that $c = \text{ord}(\beta)$ does not divide t. Thus there exists some prime p and integer $d \geq 1$ such that p^d divides c but not t. Let e be the largest power of p that divides t, so $0 \leq e < d$. Consider $\alpha = \lambda^{p^e}$, and let $k = \text{ord}(\alpha)$; then $k = t/p^e$, and $\gcd(k,p) = 1$ Consider also $\gamma = \beta^{c/p^d}$, and let $b = \text{ord}(\gamma)$; then $b = p^d$, and $\gcd(b,k) = 1$.

We now prove that $\alpha\gamma$ is an element of order $bk > t$ which contradicts the statement that t is the largest order of any element. Suppose $\text{ord}(\alpha\gamma) = s$. Then $(\alpha\gamma)^s = 1$. Since $(\alpha\gamma)^{bk} = 1$, it can be shown in an analogous fashion to Lemma 2.5 that s divides bk. We proceed to show that bk divides s, implying $s = bk$. Now

$$(\alpha\gamma)^{sb} = \alpha^{sb} = 1 \quad \text{and} \quad (\alpha\gamma)^{sk} = \gamma^{sk} = 1$$

and so k divides sb and b divides sk, as α has order k and γ has order b. Since $\gcd(b,k) = 1$, k divides s and b divides s implying bk divides s. This establishes that $s = bk$. We conclude $t = q-1$, and the proof is complete. □

It follows from the definition that a primitive element α for a field $GF(q)$ has order $q-1$. Hence $\alpha^{q-1}=\alpha^0=1$, and we see that exponent arithmetic is computed modulo $q-1$.

The following lemma is often of use in dealing with finite fields.

Lemma 2.7. If $\alpha, \beta \in F$ and F has characteristic p, then

$$(\alpha + \beta)^p = \alpha^p + \beta^p$$

Proof.

By the binomial theorem

$$(\alpha+\beta)^p = \sum_{i=0}^{p} \binom{p}{i} \alpha^{p-i}\beta^i = \binom{p}{0}\alpha^p + \sum_{i=1}^{p-1} \binom{p}{i} \alpha^{p-i}\beta^i + \binom{p}{p}\beta^p$$

Now $\binom{p}{i}$ is an integer, and for $1 \le i \le p-1$,

$$\binom{p}{i} = \frac{p(p-1)...(p-i+1)}{i(i-1)...2 \cdot 1}.$$

Since p exceeds all factors in the denominator, and p is prime, p in the numerator goes uncancelled. As a result, $\binom{p}{i} \equiv 0 \pmod{p}$ for $1 \le i \le p-1$. Hence all intermediate terms vanish, and the result follows. □

As a corollary, it is easy to see that $(\alpha+\beta+\gamma)^p = \alpha^p + \beta^p + \gamma^p$, and $(\alpha+\beta)^{p^m} = \alpha^{p^m} + \beta^{p^m}$. We say that exponentiation by a power of the characteristic p is a *linear operation* over the field.

2.4 The Minimal Polynomial

The *minimal polynomial* of a field element shall prove to be useful in §5.8, when we consider factoring $x^n - 1$ over $GF(q)$, and in Chapter 6, where we consider *BCH-codes*.

Definition. Let F be a field of characteristic p and let $\alpha \in F^*$ be a non-zero element. A *minimal polynomial* of α with respect to $GF(p)$ is a monic polynomial $m(x)$ of least degree in $GF(p)[x]$ such that $m(\alpha) = 0$.

Theorem 2.8. The minimal polynomial of an element α is unique.

Proof.

Suppose $F = GF(q)$ and F has characteristic p. It follows from Lemma 2.5 that α satisfies the polynomial $x^{q-1} - 1$ in $GF(p)[x]$. Since there is some polynomial in $GF(p)[x]$ for which α is a root, there must be one of least degree. This establishes that a minimal polynomial $m(x)$ exists. Suppose there are two monic polynomials $m_1(x)$ and $m_2(x)$ of least degree having α as a root. By the division algorithm for polynomials

$$m_1(x) = l(x)m_2(x) + r(x) \quad \text{where} \quad deg\ r(x) < deg\ m_2(x) \ \text{ or } \ r(x) = 0.$$

Since $m_1(\alpha) = 0$ and $m_2(\alpha) = 0$ we have $r(\alpha) = 0$. But $m_2(x)$ has least degree which implies $r(x) = 0$, and hence $m_2(x)$ divides $m_1(x)$. Similarly, we can show that $m_1(x)$ divides $m_2(x)$ and, since $m_1(x)$ and $m_2(x)$ are monic, $m_1(x) = m_2(x)$. This completes the proof. □

Because of the preceding result we talk about *the* minimal polynomial of an element α. We denote it $m_\alpha(x)$. For the remainder of this section, let F be a finite field of characteristic p, and recall the notation $F^* = F \backslash \{0\}$. We first note that the minimal polynomial of an element is an irreducible polynomial.

Theorem 2.9. For $\alpha \in F^*$, the minimal polynomial of α, $m_\alpha(x)$, is an irreducible polynomial.

Proof.

If $m_\alpha(x)$ is reducible, then

$$m_\alpha(x) = h(x)l(x)$$

for some $h(x)$, $l(x) \in GF(p)[x]$ having $deg\ h(x) \geq 1$ and $deg\ l(x) \geq 1$. Hence, $m_\alpha(\alpha) = h(\alpha)l(\alpha) = 0$ implies that at least one of $h(x)$ or $l(x)$ has α as a root. This contradicts the minimality of degree in the definition of $m_\alpha(x)$. □

We now proceed to show that the minimal polynomial, with respect to $GF(p)$, of an element $\alpha \in GF(p^n)$, is a polynomial in $GF(p)[x]$. First, we make use of the following definition.

Definition. For $\alpha \in F$, let t be the smallest positive integer such that $\alpha^{p^t} = \alpha$. (Note that such a t exists by Lemma 2.4.) Then the set of *conjugates* of α (with respect to $GF(p)$) is

$$C(\alpha) = \{\alpha, \alpha^p, \alpha^{p^2}, \alpha^{p^3}, \ldots, \alpha^{p^{t-1}}\}.$$

It is not difficult to see that $C(\alpha) = C(\alpha^{p^i})$ for all i, for a field F with characteristic p.

Lemma 2.10. Let F be a finite field of characteristic p, let $\alpha \in F^*$, and let $C(\alpha)$ be the set of conjugates of α with respect to $GF(p)$. Then

$$m(x) = \prod_{\beta \in C(\alpha)} (x - \beta)$$

is a polynomial with coefficients from $GF(p)$.

Proof.

Let $m(x) = \sum_{i=0}^{t} m_i x^i$. The coefficients m_i are in F; we need to prove that they are in fact in the ground field $GF(p)$. First note that

$$m(x)^p = \prod_{\beta \in C(\alpha)} (x - \beta)^p = \prod_{\beta \in C(\alpha)} (x^p - \beta^p)$$

$$= \prod_{\beta \in C(\alpha)} (x^p - \beta) = m(x^p) = \sum_{i=0}^{t} m_i x^{ip},$$

with the second equality following from Lemma 2.7, and the third since

$$\{\beta : \beta \in C(\alpha)\} = \{\beta^p : \beta \in C(\alpha)\}.$$

On the other hand, note that

$$m(x)^p = \sum_{i=0}^{t} (m_i x^i)^p = \sum_{i=0}^{t} m_i^p x^{ip}.$$

Hence $m_i = m_i^p$, implying by Lemma 2.4 that $m_i \in GF(p)$ for $0 \le i \le t$. This completes the proof. □

Suppose $m_\alpha(x) = \sum_{i=0}^{t} m_i x^i$ where $m_i \in GF(p)$. By definition, α is a root of this polynomial. But now, note that α^p is also a root, as

$$m_\alpha(\alpha^p) = \sum_{i=0}^{t} m_i \alpha^{pi} = \sum_{i=0}^{t} m_i^p \alpha^{pi} \quad (m_i^p = m_i, \text{ since } m_i \in GF(p))$$

$$= [\sum_{i=0}^{t} m_i \alpha^i]^p \quad \text{(by Lemma 2.7)}$$

$$= [m_\alpha(\alpha)]^p = 0$$

since $m_\alpha(\alpha) = 0$. It follows that all elements in $C(\alpha)$ are roots of $m_\alpha(x)$. This result, together with Lemma 2.10, provides a proof of the following theorem.

Theorem 2.11. For $\alpha \in F^*$, the minimal polynomial of α is given by

$$m_\alpha(x) = \prod_{\beta \in C(\alpha)} (x-\beta).$$

In the next section we illustrate these concepts by example.

2.5 Examples of Finite Fields

In §2.1, we constructed multiplication tables for finite fields with 4 and 9 elements. Here, we consider a few more examples of finite fields and determine for each element its minimal polynomial.

Example 5.

We construct a field $F = GF(2^3)$. First, we need an irreducible cubic polynomial over Z_2. Take $f(x) = x^3 + x + 1$. The elements of F are [0], [1], $[x]$, $[1+x]$, $[x+x^2]$, $[x^2]$, $[1+x^2]$, $[1+x+x^2]$. For simplicity, we drop the brackets which formally denote equivalence classes. Multiplication of elements is modulo the polynomial $f(x)$. Since $x^3 + x + 1 \equiv 0 \pmod{f(x)}$ we have that $x^3 \equiv -x - 1 = x + 1 \pmod{f(x)}$, since $1 = -1$ in Z_2. Thus for example the product of the field elements $1+x^2$ and $1+x$ is

$$1 + x^2 + x + x^3 = 1 + x^2 + x + (x+1) = x^2.$$

We find it convenient to write $(a_0 a_1 a_2)$ for the field element $a_0 + a_1 x + a_2 x^2$. Note that we have adopted the convention of ordering coefficients from low order to high order. Using this notation, the elements of F are

$$\begin{array}{rr} 0 = (000) & x^2 = (001) \\ 1 = (100) & 1+x^2 = (101) \\ x = (010) & x+x^2 = (011) \\ 1+x = (110) & 1+x+x^2 = (111) \end{array}$$

If we take $\alpha = x$, then it is easily checked that α is a generator of F. In fact, in this field every non-zero element other than 1 is a generator. This is not the case in every finite field (recall example 3). Suppose we take $\beta = (101)$ and compute $m_\beta(x)$. Perhaps it would be better to use the indeterminate y instead of x for the minimal polynomial so as to avoid confusion with the polynomial in x used to define the field. By Theorem 2.11,

$$m_\beta(y) = \prod_{\delta \in C(\beta)} (y-\delta) = (y-\beta)(y-\beta^2)(y-\beta^4)$$

since $\beta^8 = \beta$. We compute

$$(y-\beta)(y-\beta^2)(y-\beta^4) = y^3 + (\beta+\beta^2+\beta^4)y^2 + (\beta\beta^2+\beta\beta^4+\beta^2\beta^4)y + \beta\beta^2\beta^4.$$

To make multiplication easier, we determine the representation of each non-zero element as a power of the generator α, in this case using $\alpha = x$.

$$\begin{array}{ll} \alpha^0 = (100) & \alpha^4 = (011) \\ \alpha^1 = (010) & \alpha^5 = (111) \\ \alpha^2 = (001) & \alpha^6 = (101) \\ \alpha^3 = (110) & \alpha^7 = \alpha^0 = 1. \end{array}$$

Since $\beta = \alpha^6$, then $\beta^2 = \alpha^{12} = \alpha^5$ and $\beta^4 = \alpha^{24} = \alpha^3$. Note that exponent arithmetic is done modulo $8-1 = 7$, the order of the cyclic group generated by α. Hence

$$\begin{aligned} \beta + \beta^2 + \beta^4 &= \alpha^6 + \alpha^5 + \alpha^3 \\ &= (101) + (111) + (110) = (100) = 1 \\ \beta\beta^2 + \beta\beta^4 + \beta^2\beta^4 &= \beta^3 + \beta^5 + \beta^6 \\ &= \alpha^{18} + \alpha^{30} + \alpha^{36} \\ &= \alpha^4 + \alpha^2 + \alpha = 0 \\ \beta\beta^2\beta^4 &= \beta^7 = \alpha^{42} = 1. \end{aligned}$$

Therefore

$$m_\beta(y) = y^3 + y^2 + 1.$$

This is also the minimal polynomial of β^2 and β^4. The minimal polynomial of α is also the minimal polynomial of α^2 and α^4. It is

$$m_\alpha(y) = y^3 + y + 1.$$

Example 6.

Consider constructing $F = GF(2^4)$. Take $f(x) = x^4 + x + 1 \in Z_2[x]$. In this case $\alpha = x$ is a generator of the field.

$$\begin{array}{ll}
\alpha^0 = (1000) & \alpha^8 = (1010) \\
\alpha^1 = (0100) & \alpha^9 = (0101) \\
\alpha^2 = (0010) & \alpha^{10} = (1110) \\
\alpha^3 = (0001) & \alpha^{11} = (0111) \\
\alpha^4 = (1100) & \alpha^{12} = (1111) \\
\alpha^5 = (0110) & \alpha^{13} = (1011) \\
\alpha^6 = (0011) & \alpha^{14} = (1001) \\
\alpha^7 = (1101) & \alpha^{15} = \alpha^0 = 1.
\end{array}$$

The minimal polynomials of the elements are as follows. Let $m_i(y)$ denote $m_{\alpha^i}(y)$. Then

$$m_1(y) = y^4 + y + 1 = m_2(y) = m_4(y) = m_8(y)$$

$$m_3(y) = y^4 + y^3 + y^2 + y + 1 = m_6(y) = m_{12}(y) = m_9(y)$$

$$m_5(y) = y^2 + y + 1 = m_{10}(y)$$

$$m_7(y) = y^4 + y^3 + 1 = m_{14}(y) = m_{13}(y) = m_{11}(y)$$

$$m_0(y) = y + 1.$$

Example 7.

We will again construct $GF(2^4)$ but this time we will use the irreducible polynomial $f(x) = x^4 + x^3 + x^2 + x + 1$. In this case $\alpha = x$ is not a generator of F but $\alpha = x + 1$ is.

$$\begin{array}{ll} \alpha^0 = (1000) & \alpha^8 = (1001) \\ \alpha^1 = (1100) & \alpha^9 = (0010) \\ \alpha^2 = (1010) & \alpha^{10} = (0011) \\ \alpha^3 = (1111) & \alpha^{11} = (1101) \\ \alpha^4 = (0111) & \alpha^{12} = (0100) \\ \alpha^5 = (1011) & \alpha^{13} = (0110) \\ \alpha^6 = (0001) & \alpha^{14} = (0101) \\ \alpha^7 = (1110) & \alpha^{15} = \alpha^0 = 1. \end{array}$$

The minimal polynomials are

$$m_7(y) = y^4 + y + 1$$

$$m_3(y) = y^4 + y^3 + y^2 + y + 1$$

$$m_5(y) = y^2 + y + 1$$

$$m_1(y) = y^4 + y^3 + 1.$$

Notice that the minimal polynomials of example 7 are the same as in example 6, although they have "different names". This is not a coincidence. The two fields are *isomorphic*. They are identical from an algebraic point of view, but differ in the matter of the naming of elements.

Example 8.

We construct $F = GF(3^2)$ using $f(x) = x^2 + x + 2$. In this case $\alpha = x$ is a generator and we have

$$\begin{array}{lll} \alpha^0 = (10) & \alpha^3 = (22) & \alpha^6 = (21) \\ \alpha^1 = (01) & \alpha^4 = (20) & \alpha^7 = (11) \\ \alpha^2 = (12) & \alpha^5 = (02) & \alpha^8 = \alpha^0 = 1. \end{array}$$

Note that over Z_3, $-1 = 2$. The minimal polynomials are

$$m_1(y) = (y-\alpha)(y-\alpha^3) = y^2 - (\alpha^3+\alpha)y + \alpha^4 = y^2 + y + 2 = m_3(y)$$

$$m_2(y) = (y-\alpha^2)(y-\alpha^6) = y^2 - (\alpha^2+\alpha^6)y + 1 = y^2 + 1 = m_6(y)$$

$$m_4(y) = (y-\alpha^4) = y + 1$$

$$m_5(y) = (y-\alpha^5)(y-\alpha^7) = y^2 - (\alpha^5+\alpha^7)y + \alpha^4 = y^2 - y - 1 = m_7(y)$$

$$m_0(y) = y - 1.$$

2.6 Zech's Log Tables

We have seen that multiplication of field elements is easy when the elements are represented as powers of a generator. For the field $GF(q)$ with generator α, the product of α^a and α^b is $\alpha^a\alpha^b = \alpha^{a+b(\text{mod } q-1)}$. However, addition of field elements is not so easy when elements are given as powers of a generator. Conversely, whereas addition is easy when elements are represented as polynomials, multiplication is not. It would be convenient to use a single representation in which both operations are easy.

For small fields, to facilitate addition of field elements represented as powers of a generator, we set up the following table, called a *Zech's log table* . For each integer i, $0 \le i \le q-2$, we determine and tabulate the integer $j = z(i)$ such that $1+\alpha^i = \alpha^{z(i)}$. Then

$$\alpha^a + \alpha^b = \alpha^a(1+\alpha^{b-a(\text{mod } q-1)}) = \alpha^a\alpha^{z(b-a)} = \alpha^{a+z(b-a)}$$

and $z(b-a)$ can be obtained from the table. We illustrate this by example.

Example 9.

Consider $GF(2^3)$ defined by the irreducible polynomial $f(x) = x^3+x+1$. A list of field elements as polynomials, and their corresponding representations as powers of the generator $\alpha = x$, is given in example 5. Using this we can easily determine for each i, $0 \le i \le 6$, the integer $j = z(i)$ such that

$$1 + \alpha^i = \alpha^{z(i)}.$$

We get

$$1 + \alpha^0 = 0$$
$$1 + \alpha^1 = \alpha^3$$
$$1 + \alpha^2 = \alpha^6$$
$$1 + \alpha^3 = \alpha^1$$
$$1 + \alpha^4 = \alpha^5$$
$$1 + \alpha^5 = \alpha^4$$
$$1 + \alpha^6 = \alpha^2$$

For example, $1 + \alpha^3 = (100) + (110) = (010) = \alpha^1$. Defining $\alpha^\infty = 0$, we summarize this information in the following Zech's log table.

i	$z(i)$
∞	0
0	∞
1	3
2	6
3	1
4	5
5	4
6	2

With this table, representing elements of $F = GF(2^3)$ as powers of the generator $\alpha = x$, both multiplication and addition are easily performed. For multiplication, we simply add exponents and reduce modulo $q-1 = 7$, so for example

$$\alpha^4\alpha^6 = \alpha^{10 (\text{mod } 7)} = \alpha^3.$$

As an example of addition, to determine $\alpha^3 + \alpha^5$ as a power of α we compute

$$\alpha^3 + \alpha^5 = \alpha^3(1+\alpha^2) = \alpha^3(\alpha^6) = \alpha^9 = \alpha^2,$$

using the Zech's log table to look up $1+\alpha^2 = \alpha^6$.

Example 10.

Consider the field $F=GF(2^4)$ generated by the root $\alpha = x$ of the irreducible polynomial $f(x) = x^4+x+1 \in Z_2[x]$, as in example 6. The Zech's log table for this field is given further below. The table can easily be constructed once the representations of all field elements as both polynomials and as powers of a generator are known. Alternatively, the table can often be constructed more rapidly in an ad hoc manner, illustrated as follows. Since α is a root of x^4+x+1 and since + and – are equivalent in Z_2 we have $\alpha^4+\alpha+1 = 0$ and

$$1 + \alpha = \alpha^4. \tag{1}$$

This equation also implies $1+\alpha^4 = \alpha$, providing table entries $(i, z(i))=(1,4)$ and $(4,1)$. Squaring both sides of equation (1) and using Lemma 2.7 (with p=2) produces the relation

$$1 + \alpha^2 = \alpha^8 \tag{2}$$

which yields entries (2,8) and (8,2). Now squaring both sides of (2) returns us to relation

(1), giving nothing new. However, multiplying through (1) by α^{-1} gives

$$\alpha^{14} + 1 = \alpha^3 \tag{3}$$

from which we derive entries (14,3) and (3,14). Squaring both sides of (3) provides entries (13,6) and (6,13). In a similar manner, multiplying through (1) by α^{-4} provides entries (11,12), (12,11), (7,9) and (9,7). The entries remaining to complete the table are those for i=5 and i=10. Since (5,5) and (10,10) clearly do not satisfy the Zech's relation, the only remaining possibility is the set of pairings (5,10) and (10,5).

$1+\alpha^i = \alpha^{z(i)}$				
i	$z(i)$		i	$z(i)$
∞	0		7	9
0	∞		8	2
1	4		9	7
2	8		10	5
3	14		11	12
4	1		12	11
5	10		13	6
6	13		14	3

2.7 Exercises

1. Determine whether or not finite fields exist with the following number of elements:
(a) 2 (b) 3 (c) 17 (d) 21 (e) 27 (f) 64

2. Determine the characteristic of each of the following fields.
(a) $GF(2)$ (b) $GF(8)$ (c) $GF(29)$ (d) $GF(81)$ (e) $GF(125)$

3. Is Z_2 isomorphic to $GF(2)$? Z_4 to $GF(4)$? Z_7 to $GF(7)$? Z_9 to $GF(9)$?

4. (a) Verify that $f(x) = x^5 - 1 \in GF(31)[x]$ can be written as the product $(x^2-3x+2)(x^3+3x^2+7x+15)$.
(b) Determine the congruence class of $f(x)$ modulo (x^2-3x+2).
(c) Compute $f(x) \pmod{x^5}$; $x^7 \pmod{f(x)}$; and $x^6+5x^3 \pmod{f(x)}$.

5. Let $I(p,n)$ be the number of irreducible monic polynomials of degree n over $GF(p)$.

   (a) Derive an expression for $I(p,2)$, i.e. the number of polynomials $x^2 + bx + c$, where $b, c \in GF(p)$, which are irreducible over $GF(p)$.

   (b) Using this expression, write down an unsimplified expression for $I(p,3)$, the number of irreducible cubics $x^3 + bx^2 + cx + d$.

   (c) Write down an unsimplified expression for $I(p,4)$.

6. All linear polynomials are irreducible by definition. All quadratic and cubic polynomials are either irreducible, or have a linear factor, and the latter can be ascertained by using the *factor theorem*, which states that $f(x) \in F[x]$ has a linear factor $(x-\alpha)$ if and only if $f(\alpha) = 0$. A quartic polynomial is either irreducible, has a linear factor, or is the product of two irreducible quadratics.
(a) Find all irreducible polynomials of degrees 2, 3 and 4 over $GF(2)$.
(b) Find all irreducible polynomials of degrees 2 and 3 over $GF(3)$.

7. The number of monic irreducible polynomials of degree n over the field $GF(q)$ is given by the formula $I(q,n) = (1/n)\sum_{c|n}\mu(n/c)q^c$, where the sum is taken over all distinct positive divisors c of n, and $\mu(n)$ is the *Möbius μ-function*, defined on the positive integers by $\mu(1) = 1$, $\mu(c) = 0$ if $c = a^2b$ for any $a > 1$, and $\mu(c) = (-1)^k$ if $c = p_1p_2 \ldots p_k$, where the p_i are distinct primes.
(a) Determine the number of monic irreducible polynomials of degrees 2, 3 and 4 over $GF(2)$.
(b) Determine the number of monic irreducible polynomials of degrees 2, 3 and 4 over $GF(3)$.

8. (a) Construct the addition and multiplication tables for a field with 8 elements; represent the elements as polynomials. (Hint: verify that $f(x)=x^3+x^2+1$ is an irreducible polynomial over $GF(2)$)

   (b) Find a generator in the field.

9. (a) Verify that $\alpha=x$ is a generator for the field with 8 elements constructed using the irreducible polynomial $f(x)=x^3+x^2+1$ over $GF(2)$.

   (b) Represent each element as a power of the generator α, and give the corresponding polynomial representation of each element.

   (c) Construct the multiplication table for the field, representing the elements as polynomials within the table, but using the correspondence from (b) to simplify the multiplication operation.

   (d) Construct a Zech's log table for the field.

   (e) Find the minimum polynomial of each field element.

10. Consider the field with 16 elements constructed using the irreducible polynomial $f(x)=x^4+x^3+1$ over $GF(2)$.

    (a) Verify that $f(x)$ is irreducible over $GF(2)[x]$.

    (b) Show that $\alpha=x$ is a generator for the field, and represent each element as both a polynomial and as a power of the generator.

    (c) Construct a Zech's log table for the field.

    (d) Find the minimum polynomial of each field element.

11. In the field with 16 elements generated by the root α of the irreducible polynomial $f(x)=x^4+x+1$ over $GF(2)$, find the order of the following elements:

    (a) 1 (b) $\alpha^1=x$ (c) $\alpha^4=1+x$ (d) $\alpha^5=x+x^2$ (e) $\alpha^6=x^2+x^3$

12. (a) Verify that $f(x)=x^2+x+2$ is irreducible over $GF(3)$.

    (b) Construct a Zech's log table for the field with 9 elements generated by $\alpha=x$.

13. Let $f(x)\in GF(p)[x]$. Prove that congruence (modulo $f(x)$) of polynomials in $GF(p)[x]$ is an equivalence relation. That is, for all $a(x), b(x), c(x)\in GF(p)[x]$, prove

    (a) $a(x)\equiv a(x)$

    (b) if $a(x)\equiv b(x)$ then $b(x)\equiv a(x)$

    (c) if $a(x)\equiv b(x)$ and $b(x)\equiv c(x)$, then $a(x)\equiv c(x)$

14. Consider the Zech's log table for the finite field $GF(2^5)$ given in Appendix D, with generator α.

 (a) Determine β as a power of α, where $\beta = \alpha^5 + \alpha^{23} + \dfrac{\alpha^2 + \alpha^4}{1 + \alpha^{12}}$.

 (b) Find the two roots of the quadratic equation $\alpha^3 x^2 + \alpha^{18} x + 1 = 0$ in $GF(2^5)$. (Hint: Let $x = (\alpha^{18}/\alpha^3)y$, yielding an equation of the form $ay^2 + ay + 1 = 0 = a(y^2 + y + a^{-1})$. Then letting $y = \alpha^i$, observe $y^2 + y = y(1+y) = \alpha^i \alpha^{z(i)}$.)

15. Consider the Zech's log table for the finite field $GF(2^5)$ given in Appendix D, with generator α.

 (a) Determine δ as a power of α, where $\delta = \alpha^{10} + \dfrac{\alpha^7 + \alpha^{23}}{\alpha^4 + \alpha^9} + 1$.

 (b) Find all roots of the equation $\alpha^7 x^2 + (\alpha^3 + \alpha^9)x + \dfrac{(1+\alpha)}{\alpha^4} = 0$ in $GF(2^5)$.

16. (a) Construct the Zech's log table for $GF(2^4)$, defined using a root α of the irreducible polynomial $f(x) = x^4 + x^3 + x^2 + x + 1$.

 (b) Let $\beta = 1 + x$ be an element of this field. Find all values z in the field which satisfy $z^2 + (\beta+1)z + (1 + \beta^2 + \beta^5) = 0$.

17. Prove that a polynomial of degree n over a finite field F has at most n roots in F. (Hint: apply the division algorithm and use induction)

18. If k is a positive divisor of m then prove that $GF(p^m)$ contains a unique subfield of order p^k.

19. Determine the number of subfields in
 (i) $GF(2^{12})$, (ii) $GF(3^6)$, (iii) $GF(5^{18})$.

Chapter 3

LINEAR CODES

3.1 Introduction

In this chapter we introduce *linear codes*. The algebraic structure of linear codes provides a framework from which efficient encoding and decoding algorithms can be constructed. Many error-correcting codes in use and under investigation are subclasses of linear codes defined by imposing additional structural constraints. The cyclic codes of Chapter 5 are a prime example, and the BCH-codes of Chapter 6 are a further refinement of these.

We suppose that the set of all messages that we wish to transmit is the set of k-tuples having components from some field F with q elements. We denote the set of all k-tuples over F by $V_k(F)$. Then there are q^k messages. This set is a *vector space*. (For a review of vector spaces, see Appendix A.) We refer to it as the "message space". For example, if the letters of the alphabet are assigned to a subset of the set of all binary 5-tuples, then we can transmit letters of the alphabet as 5-tuples. Our message space consists of all 2^5 binary 5-tuples. Of course, we only use 26 of these.

In order to detect and possibly correct errors, we will have to add some redundancy. Hence, the message k-tuples will be embedded into n-tuples, $n \geq k$. Thus, we will set up a one-to-one correspondence between the q^k messages and q^k n-tuples in $V_n(F)$. An algebraically convenient technique is to make sure that the q^k n-tuples form a k-dimensional *subspace* in $V_n(F)$. We recall that a k-dimensional subspace is determined by k linearly independent vectors in $V_n(F)$. There are many such k-dimensional subspaces from which to choose. This number can be easily enumerated as follows.

We can select k linearly independent vectors in

$$(q^n-1)(q^n-q)(q^n-q^2)\cdots(q^n-q^{k-1}) \tag{3.1}$$

ways. The $q^n - 1$ comes from the fact that we can select any non-zero vector. The (q^n-q) term appears since we can select any vector which is linearly independent from the first vector chosen. There are precisely q vectors which are linearly dependent on a given vector. In general, if i vectors are linearly independent, they generate a subspace of dimension i which contains q^i vectors. In order to select a vector which is independent from these i vectors, we have exactly $q^n - q^i$ choices. Thus, (3.1) follows. Expression (3.1) is not the number of k-dimensional subspaces since it just counts the number of ordered sets of k linearly independent vectors. We need to know how many sets of k linearly independent vectors generate the same k-dimensional subspace. By reasoning similar to that used in determining (3.1), we get that a k-dimensional subspace contains precisely

$$(q^k-1)(q^k-q)\cdots(q^k-q^{k-1}) \tag{3.2}$$

ordered sets of k linearly independent vectors. Therefore, the total number of k-dimensional subspaces in $V_n(F)$ is

$$s = \frac{(q^n-1)(q^n-q)\cdots(q^n-q^{k-1})}{(q^k-1)(q^k-q)\cdots(q^k-q^{k-1})} = \frac{(q^n-1)(q^{n-1}-1)\cdots(q^{n-k+1}-1)}{(q^k-1)(q^{k-1}-1)\cdots(q-1)}.$$

Hence there are many k-dimensional subspaces of $V_n(F)$ from which to choose.

Example 1.

Consider the number s of subspaces of dimension 3 in $V_5(Z_2)$. We have $k=3$, $n=5$, $q=2$ and

$$s = \frac{(2^5-1)(2^4-1)(2^3-1)}{(2^3-1)(2^2-1)(2-1)} = 155.$$

The number of subspaces of dimension 2 in $V_5(Z_2)$ is also 155.

If we select any one of these s subspaces, we can define a one-to-one correspondence between the vectors in this subspace S and the message space M. The most convenient method is to select first a *basis* $B = \{\mathbf{v}_1, \mathbf{v}_2, \ldots, \mathbf{v}_k\}$ for S and then define the correspondence

$$f: M \to S$$

by

$$f(\mathbf{m}) = \sum_{i=1}^{k} m_i \mathbf{v}_i$$

where $\mathbf{m} = (m_1 m_2 \ldots m_k)$ is a message k-tuple, i.e. $\mathbf{m} \in M$.

Example 2.

Consider a message space M of 4 message 2-tuples,

$$M = \{(00), (10), (01), (11)\}.$$

We select a subspace S of $V_4(Z_2)$ of dimension $k = 2$ defined by the basis $B = \{\mathbf{v}_1, \mathbf{v}_2\}$, where

$$\mathbf{v}_1 = (1100)$$

$$\mathbf{v}_2 = (0110)$$

Then f maps M to S as follows.

$$\begin{aligned} (0\,0) &\rightarrow (0\,0\,0\,0) \\ (1\,0) &\rightarrow (1\,1\,0\,0) \\ (0\,1) &\rightarrow (0\,1\,1\,0) \\ (1\,1) &\rightarrow (1\,0\,1\,0). \end{aligned}$$

$$S = \{(0000), (1100), (0110), (1010)\}.$$

Example 3.

Here, S is a 2-dimensional subspace of $V_5(Z_2)$. The message space is as before,

$$M = \{(00), (10), (01), (11)\}.$$

If the basis for S is chosen to be $B = \{\mathbf{v}_1, \mathbf{v}_2\}$ where

$$\mathbf{v}_1 = (10111)$$

$$\mathbf{v}_2 = (11110)$$

then the mapping is

$$\begin{aligned} (00) &\rightarrow (00000) \\ (10) &\rightarrow (10111) \\ (01) &\rightarrow (11110) \\ (11) &\rightarrow (01001) \end{aligned}$$

and

$$S = \{(00000), (10111), (01001), (11110)\} .$$

So far, we have not been too concerned with the value of n and the particular k-dimensional subspace of $V_n(F)$ selected for the correspondence. For fixed k, the larger n is, the more *check digits* there are, giving greater potential for error correction. Hence for fixed k, as mentioned earlier, choosing n typically involves a tradeoff between error-correcting capability and the rate of the code. However, once k and n are both fixed, note that not all k-dimensional subspaces have the same Hamming distance, which we saw in §1.3 affects the number of errors that can be corrected. Note that the subspace S of example 3 above has distance $d = 2$. It is easy to verify that had we chosen the basis $\mathbf{v}_1 = (10000)$, $\mathbf{v}_2 = (01000)$, then the resulting subspace would have had distance 1, and had we chosen $\mathbf{v}_1 = (10110)$, $\mathbf{v}_2 = (01011)$, then the resulting subspace would have had distance 3. Here we should simply choose a subspace which provides the greatest Hamming distance. As the choice of a k-dimensional subspace is important, a study of subspaces with respect to Hamming distance is necessary. This leads to our definition of a *linear code*.

Definition. A *linear (n,k)-code* over F is a k-dimensional subspace of $V_n(F)$.

We reserve round brackets to denote such codes, and hence often omit the word "linear". We sometimes refer to a linear code as an (n,k,d)-code, to include specification of its distance d. In Chapter 1 we observed that in order to compute the Hamming distance in an arbitrary code of length n having t codewords we must check all $\binom{t}{2}$ pairs of codewords. For linear codes, we shall see that the work is reduced to calculating a related parameter, the ***Hamming weight*** of the code. We make the following definitions, analogous to those for Hamming distance in §1.3.

Definition. The *Hamming weight of a vector* $\mathbf{v} \in V_n(F)$, denoted $w(\mathbf{v})$, is the number of non-zero coordinates in $\mathbf{v}$.

Definition. The *Hamming weight of an (n,k)-code C* is

$$w(C) = \min\{w(\mathbf{x}): \mathbf{x} \in C, \mathbf{x} \neq 0\}.$$

Example 4.

Consider

$$S_1 = \{(0000), (1000), (0100), (1100)\}$$

$$S_2 = \{(0000), (1100), (0011), (1111)\}.$$

S_1 and S_2 are both 2-dimensional subspaces of $V_4(Z_2)$. The Hamming distance and weight of S_1 are both 1, whereas for S_2 they are both 2.

In the example above, note that the Hamming distance of each code is equal to the weight of that code. This is not a coincidence.

Theorem 3.1. Let d be the distance in an (n,k)-code C. Then

$$d = w(C).$$

Proof.

By definition, the distance of the code C is given by

$$d = \min\{d(\mathbf{x},\mathbf{y}):\ \mathbf{x},\mathbf{y} \in C, \mathbf{x} \neq \mathbf{y}\}.$$

It is easy to see that $d(\mathbf{x},\mathbf{y}) = w(\mathbf{x}-\mathbf{y})$ since component i of $\mathbf{x}-\mathbf{y}$ is non-zero if and only if components i of $\mathbf{x}$ and $\mathbf{y}$, which are elements of the field F, differ (see Chapter 1, exercise 12(a)). Since C is a linear code, and subspaces are closed under addition, $\mathbf{x} - \mathbf{y} \in C$. Therefore

$$d = \min\{w(\mathbf{z}): \mathbf{z} \in C, \mathbf{z} \neq 0\} = w(C).$$

□

Hence computing the distance of a linear code reduces to computing the Hamming weight of the code, which is the weight of the "lightest" non-zero codeword.

Let C be an (n,k)-code. Since C is a k-dimensional subspace, C is completely determined by k linearly independent vectors. As noted earlier, once such a basis is selected we can set up a one-to-one correspondence in a natural way between the k-dimensional message space and the code C. Since we can select

$$t = (q^k-1)(q^k-q)\ldots(q^k-q^{k-1})$$

ordered sets of k linearly independent vectors in C, there are t one-to-one correspondences we can establish between the message space M and a fixed k-dimensional subspace C. (There are, of course, many others but these mappings corresponding to the different bases are the most natural.) For a fixed k-dimensional subspace C, there are advantages in selecting one particular basis over another.

Example 5.

The vectors $\mathbf{v}_1 = (10000)$, $\mathbf{v}_2 = (11010)$, and $\mathbf{v}_3 = (11101)$ form a basis for a (5,3)-code C over Z_2. The code C consists of the vectors

$$(00000), (10000), (11010), (11101), (01010), (01101), (00111), (10111).$$

For convenience, write

$$G = \begin{bmatrix} \mathbf{v}_1 \\ \mathbf{v}_2 \\ \mathbf{v}_3 \end{bmatrix} = \begin{bmatrix} 1 & 0 & 0 & 0 & 0 \\ 1 & 1 & 0 & 1 & 0 \\ 1 & 1 & 1 & 0 & 1 \end{bmatrix}.$$

If $\mathbf{m} = (m_1m_2m_3) \in M$, the message space of binary 3-tuples, then the codeword in C associated with $\mathbf{m}$ is

$$\mathbf{m}G = m_1\mathbf{v}_1 + m_2\mathbf{v}_2 + m_3\mathbf{v}_3.$$

For example, if $\mathbf{m} = (101)$, then

$$\mathbf{m}G = (1\ 0\ 1) \begin{bmatrix} 1 & 0 & 0 & 0 & 0 \\ 1 & 1 & 0 & 1 & 0 \\ 1 & 1 & 1 & 0 & 1 \end{bmatrix} = (0\ 1\ 1\ 0\ 1).$$

Suppose that instead of $\mathbf{v}_1$, $\mathbf{v}_2$, $\mathbf{v}_3$, we select the basis $\mathbf{u}_1 = (10000)$, $\mathbf{u}_2 = (01010)$, $\mathbf{u}_3 = (00111)$ for C (note that this basis does indeed generate the same 3-dimensional subspace), and let

$$G' = \begin{bmatrix} \mathbf{u}_1 \\ \mathbf{u}_2 \\ \mathbf{u}_3 \end{bmatrix} = \begin{bmatrix} 1 & 0 & 0 & 0 & 0 \\ 0 & 1 & 0 & 1 & 0 \\ 0 & 0 & 1 & 1 & 1 \end{bmatrix}.$$

Then, using G', $\mathbf{m} = (101)$ gets mapped to the codeword (10111). In this codeword, the first three positions are precisely the *information symbols* or message. Therefore to retrieve the message, one needs only to take the first three components of the codeword. The basis $\mathbf{v}_1, \mathbf{v}_2, \mathbf{v}_3$ does not have this property and hence decoding is more complicated.

3.2 Generator Matrices and the Dual Code

As illustrated in example 5, a linear code can be characterized by a *generator matrix*.

Definition. A *generator matrix* G for an (n,k)-code C is a $k \times n$ matrix whose rows are a vector space basis for C.

The codewords of a linear code C (over the field F) with generator matrix G are all linear combinations (over F) of the rows of G, i.e. those vectors in the *row space* of G. We say that C is the code *generated* by the matrix G. In addition, we may associate with any $k' \times n$ matrix G' (over F) the (n,k)-code C (over F) consisting of those vectors in the row space of G'. In this case, $k \leq k'$. If some rows of G' are linearly dependent then $k < k'$, and a generator matrix for C fitting the definition above can be derived by the removal from G' of the rows causing the dependencies. We may on occasion speak of such a matrix G' as a generator matrix for C, although this is not technically correct when $k < k'$; however, the meaning should be clear. In either case, the dimension k of the associated code is the number of linearly independent rows in the defining matrix, i.e. the *rank* of the matrix.

Since the basis for a k-dimensional vector space is not unique, neither is the generator matrix G for a linear code C. *Elementary row operations* (multiplying a row by a scalar, adding a scalar multiple of one row to another, and exchanging two rows) performed on G give matrices which also generate C. As the previous example illustrates, one generator matrix may be more useful than another. For example, if we can find a generator matrix having the form

$$G = [I_k A]$$

where I_k is the $k \times k$ identity matrix and A is a $k \times (n-k)$ matrix, then the information symbols will occur in the first k positions of a codeword. Such a matrix G is said to be in *standard form.*

Unfortunately, we cannot always guarantee that such a G will exist for C. However, we shall see that this problem is easily overcome. It is not hard to see that permuting the coordinate positions of a code C gives a subspace C' having precisely the same Hamming weight and distance. In this sense, C and C' are essentially the same code. This motivates the following definition. Recall that a *permutation matrix* is an identity matrix with rows or columns permuted.

Definition. Two (n,k)-codes C and C' over a field F are said to be *equivalent codes* if there exist generator matrices G and G' for C and C' respectively and an $n \times n$ permutation matrix P such that

$$G' = GP.$$

The matrix P permutes the columns of G, and thus permutes the coordinate positions in C to produce the code C'.

Example 6.

Let C be the (4,3)-code having generator

$$\tilde{G} = \begin{bmatrix} 0 & 0 & 1 & 1 \\ 0 & 1 & 1 & 0 \\ 1 & 0 & 1 & 1 \end{bmatrix}.$$

Let C' be the (4,3)-code having generator matrix

$$G' = \begin{bmatrix} 1 & 0 & 0 & 1 \\ 0 & 1 & 0 & 1 \\ 0 & 0 & 1 & 0 \end{bmatrix}.$$

We can show C and C' to be equivalent codes as follows. By row operations on $\tilde{G}$ (add row 1 to rows 2 and 3), another generating matrix for C is

$$G = \begin{bmatrix} 0 & 0 & 1 & 1 \\ 0 & 1 & 0 & 1 \\ 1 & 0 & 0 & 0 \end{bmatrix}.$$

Now, if we select the permutation matrix

$$P = \begin{bmatrix} 0 & 0 & 1 & 0 \\ 0 & 1 & 0 & 0 \\ 1 & 0 & 0 & 0 \\ 0 & 0 & 0 & 1 \end{bmatrix}$$

then $G' = GP$. P interchanges columns 1 and 3 of G, and hence interchanges coordinates 1 and 3 in each codeword of C. Thus the two codes are equivalent. Note, however, that these codes are not identical. To verify this last statement, the reader is encouraged to construct the codes explicitly using the matrices $\tilde{G}$ and G'. (Each code consists of 8 codewords.)

The definition of equivalent codes thus easily leads to the following result (exercise 8).

Theorem 3.2. If C is an (n,k)-code over a field F, then there exists a generator matrix G for C or for an equivalent code C' such that

$$G = [I_k\, A].$$

An interesting and useful way to characterize an (n,k)-code C over a field F is in terms of another subspace, called the *orthogonal complement* of C. Before defining this subspace, a preliminary definition is required.

Definition. Let $\mathbf{x} = (x_1 x_2 \ldots x_n)$ and $\mathbf{y} = (y_1 y_2 \ldots y_n)$ be vectors in $V_n(F)$. The *inner product* of $\mathbf{x}$ and $\mathbf{y}$ is

$$\mathbf{x} \cdot \mathbf{y} = \sum_{i=1}^{n} x_i y_i$$

where the sum is computed over the field F.

A few words concerning this inner product are in order. If F is the field of real numbers, then $\mathbf{x} \cdot \mathbf{x} = 0$ implies $\mathbf{x} = 0$. This need not be the case when F is a finite field. For example, if $F = Z_2$ and $\mathbf{x} = (101)$ then

$$\mathbf{x} \cdot \mathbf{x} = 1 + 0 + 1 = 0$$

yet $\mathbf{x}$ is not the zero vector.

It is easy to see from the definition that the inner product is commutative. The following properties are also easily established.

(1) For $\mathbf{x},\mathbf{y},\mathbf{z} \in V_n(F)$, $(\mathbf{x}+\mathbf{y}) \cdot \mathbf{z} = \mathbf{x} \cdot \mathbf{z} + \mathbf{y} \cdot \mathbf{z}$.

(2) For $\mathbf{x},\mathbf{y} \in V_n(F)$ and $\lambda \in F$, $(\lambda \mathbf{x}) \cdot \mathbf{y} = \lambda(\mathbf{x} \cdot \mathbf{y})$.

If $\mathbf{x} \cdot \mathbf{y} = 0$, we say that the vectors $\mathbf{x}$ and $\mathbf{y}$ are *orthogonal* to each other.

Definition. Let C be an (n,k)-code over a field F. The *orthogonal complement* $C^\perp$ ("C perp") of C is

$$C^\perp = \{\mathbf{x} \in V_n(F):\ \mathbf{x}\cdot\mathbf{y} = 0 \text{ for all } \mathbf{y} \in C\}.$$

That is, $C^\perp$ is the set of n-tuples over F that are orthogonal to every vector in C. $C^\perp$ is sometimes referred to as the *dual code* of C.

We now give a fundamental result relating a linear code and its dual.

Theorem 3.3. If C is an (n,k)-code over a field F, then $C^\perp$ is an $(n,n-k)$-code over F.

Proof.

We must prove that $C^\perp$ is an $(n-k)$-dimensional subspace of $V_n(F)$. We show that $C^\perp$ is a subspace by verifying the vector space axioms. In fact, we need only show that $C^\perp$ is closed under vector addition and under multiplication by a scalar. Most of the other axioms follow by inheritance, $C^\perp$ being a subset of $V_n(F)$. Clearly the zero vector is in $C^\perp$, and the existence of additive inverses will follow from closure under scalar multiplication (by the scalar -1).

To prove closure under vector addition, we proceed as follows. Let $\mathbf{x},\mathbf{y} \in C^\perp$. Then $\mathbf{x}\cdot\mathbf{z} = 0$ and $\mathbf{y}\cdot\mathbf{z} = 0$ for all $\mathbf{z} \in C$. Consider the vector $\mathbf{x} + \mathbf{y}$.

$$(\mathbf{x}+\mathbf{y})\cdot\mathbf{z} = \mathbf{x}\cdot\mathbf{z} + \mathbf{y}\cdot\mathbf{z} = 0$$

and thus $\mathbf{x}+\mathbf{y} \in C^\perp$. Hence $C^\perp$ is closed under vector addition. Now for $\mathbf{x} \in C^\perp$ and $\lambda \in F$, consider the scalar multiple $\lambda\mathbf{x}$. Since $\mathbf{x}\cdot\mathbf{z} = 0$ for all $\mathbf{z} \in C$,

$$(\lambda\mathbf{x})\cdot\mathbf{z} = \lambda(\mathbf{x}\cdot\mathbf{z}) = 0$$

and thus $\lambda\mathbf{x} \in C^\perp$. Therefore $C^\perp$ is closed under scalar multiplication. As argued above, it follows that $C^\perp$ is a subspace of $V_n(F)$.

To complete the proof, we now show that $C^\perp$ has dimension $n-k$. Let G be a generator matrix for C. We suppose that

$$G = [I_k\, A].$$

If no such matrix exists for C, then use an equivalent code C'. If the result is true for C', it is true for C. Consider the matrix $H = [-A^T \, I_{n-k}]$. Let $span(H)$ be the subspace generated by the rows of H, i.e. the row space of H. Now $GH^T = I_k(-A) + AI_{n-k} = 0$, implying that the rows of H are orthogonal to the rows of G, and by the simple properties of the inner product, $span(H) \subseteq C^{\perp}$. We now show that $C^{\perp} \subseteq span(H)$, which then proves that $C^{\perp} = span(H)$.

Consider any $\mathbf{x} \in C^{\perp}$ where $\mathbf{x} = (x_1 x_2 ... x_n)$ and let

$$\mathbf{y} = \mathbf{x} - \sum_{i=1}^{n-k} x_{i+k} \mathbf{r}_i$$

where $\mathbf{r}_i$ is the ith row of H. Since $\mathbf{x} \in C^{\perp}$, and we have just proven that $\mathbf{r}_i \in C^{\perp}$, $1 \leq i \leq k$, it follows by closure that $\mathbf{y} \in C^{\perp}$. We now examine the structure of $\mathbf{y}$. By construction, components k+1 through n are 0, so $\mathbf{y} = (y_1 y_2 ... y_k 0...0)$, $y_i \in F$. But since $\mathbf{y} \in C^{\perp}$, $G\mathbf{y}^T = \mathbf{0}$ which implies $y_i = 0$, $1 \leq i \leq k$. Therefore, $\mathbf{y} = \mathbf{0}$ and

$$\mathbf{x} = \sum_{i=1}^{n-k} x_{i+k} \mathbf{r}_i .$$

Hence $\mathbf{x} \in span(H)$ implying $C^{\perp} \subseteq span(H)$. Thus $C^{\perp} = span(H)$ and $C^{\perp}$ has dimension $n-k$, since the rows of H are linearly independent. This completes the proof. □

The above proof reveals a method for constructing a basis for $C^{\perp}$. We state this important construction as a corollary to Theorem 3.3.

Corollary 3.4. If $G = [I_k \, A]$ is a generator matrix for C, then $H = [-A^T \, I_{n-k}]$ is a generator matrix for $C^{\perp}$.

Example 7.

Let C be the (6,3)-code over Z_2 generated by

$$\tilde{G} = \begin{bmatrix} 1 & 0 & 1 & 1 & 0 & 1 \\ 1 & 1 & 0 & 1 & 0 & 0 \\ 0 & 1 & 0 & 0 & 1 & 1 \end{bmatrix}.$$

By applying suitable elementary row operations to $\tilde{G}$, we obtain a matrix G generating the same code, where

$$G = \begin{bmatrix} 1&0&0&1&1&1 \\ 0&1&0&0&1&1 \\ 0&0&1&0&1&0 \end{bmatrix} = [I_3\, A], \qquad A = \begin{bmatrix} 1&1&1 \\ 0&1&1 \\ 0&1&0 \end{bmatrix}.$$

A generator matrix for the orthogonal complement $C^{\perp}$ of C is

$$H = [-A^T\, I_3] = \begin{bmatrix} 1&0&0&1&0&0 \\ 1&1&1&0&1&0 \\ 1&1&0&0&0&1 \end{bmatrix}.$$

As a check, we verify that $GH^T = 0$, and note that H has $n-k = 3$ linearly independent rows.

Example 8.

Let C be the (7,4)-code over Z_3 having generator matrix

$$\tilde{G} = \begin{bmatrix} 2&1&0&1&1&0&0 \\ 0&0&1&1&0&2&0 \\ 2&1&0&2&0&2&0 \\ 1&1&0&0&0&2&1 \end{bmatrix}.$$

Subtracting row 3 from rows 1 and 4, we obtain from $\tilde{G}$ a generator matrix G for the same code C, which has among its columns those of I_4.

$$G = \begin{bmatrix} 0&0&0&2&1&1&0 \\ 0&0&1&1&0&2&0 \\ 2&1&0&2&0&2&0 \\ 2&0&0&1&0&0&1 \end{bmatrix}.$$

In order to construct a generator matrix for $C^{\perp}$, let us first construct a generator matrix G' for a code E equivalent to C such that $G' = [I_4\, A]$. This is easily done by permuting columns of G, selecting, in order, columns 5, 3, 2, 7, 1, 4 and 6 (which corresponds to post-multiplying G by the appropriate permutation matrix P).

$$G' = \begin{bmatrix} 1&0&0&0&0&2&1 \\ 0&1&0&0&0&1&2 \\ 0&0&1&0&2&2&2 \\ 0&0&0&1&2&1&0 \end{bmatrix} = [I_4\, A], \qquad A = \begin{bmatrix} 0&2&1 \\ 0&1&2 \\ 2&2&2 \\ 2&1&0 \end{bmatrix}.$$

Now, a generator matrix for $E^{\perp}$ is

$$H' = [-A^T\ I_3] = \begin{bmatrix} 0&0&1&1&1&0&0 \\ 1&2&1&2&0&1&0 \\ 2&1&1&0&0&0&1 \end{bmatrix}.$$

By rearranging the columns of H', restoring them to their original order by the selection 5, 3, 2, 6, 1, 7, 4, a generator matrix for $C^{\perp}$ is

$$H = \begin{bmatrix} 1&1&0&0&0&0&1 \\ 0&1&2&1&1&0&2 \\ 0&1&1&0&2&1&0 \end{bmatrix}.$$

As a check, note that $GH^T = 0$, and that H has the proper number ($n-k=3$) of linearly independent rows.

3.3 The Parity-Check Matrix

From the previous section, we see that a vector $\mathbf{x} \in V_n(F)$ is a codeword of an (n,k)-code C if and only if $H\mathbf{x}^T = 0$, where H is a generator matrix for $C^{\perp}$. This leads us to make the following definition.

Definition. Let C be an (n,k)-code over F. If H is a generator matrix for $C^{\perp}$, then H is called a *parity-check matrix* for C.

If G is a generator matrix for C and H is a generator matrix for $C^{\perp}$, then H is a parity-check matrix for C and G is a parity-check matrix for $C^{\perp}$. In particular, if $G = [I_k\ A]$ is a generator matrix for C, then $H = [-A^T\ I_{n-k}]$ is a parity-check matrix for C, and a generator matrix for $C^{\perp}$. It follows that we can now specify a linear code C by giving either a generator matrix G or a parity-check matrix H.

Some explanation for the choice of the term *parity-check* matrix is in order. Suppose $\mathbf{m} = (m_1 m_2 \ldots m_k)$ is a message k-tuple which we are going to embed in a codeword $\mathbf{c}$ of length n. Suppose further that the information symbols occur in the first k components of $\mathbf{c}$. Then $\mathbf{c}$ has the form

$$\mathbf{c} = (m_1\, m_2 \ldots m_k\, x_1\, x_2 \ldots x_{n-k}).$$

The x_i, $1 \le i \le n-k$, are usually referred to as the *check symbols* since they provide the necessary redundancy to possibly detect and correct errors. Since $H\mathbf{c}^T$ must be the zero vector if $\mathbf{c}$ is a codeword, given the m_i, $1 \le i \le k$, the x_i, $1 \le i \le n-k$, can be uniquely determined from the system of equations

$$H\mathbf{c}^T = 0.$$

Therefore, the check symbols x_i are determined if we are given a parity-check matrix H for the code. For the simple case $n = k+1$ (a single check symbol) in the example below, H is such that the check symbol is a *parity bit.*

Example 9.

Let $H = [1\ 1\ 1\ 1\ 1]$ be a parity-check matrix for a (5,4)-code C over Z_2. A generator matrix for C is

$$G = \begin{bmatrix} 1 & 0 & 0 & 0 & 1 \\ 0 & 1 & 0 & 0 & 1 \\ 0 & 0 & 1 & 0 & 1 \\ 0 & 0 & 0 & 1 & 1 \end{bmatrix}.$$

Notice that the information k-tuple (1011) is assigned the codeword (10111). The check bit is determined by H to be a 0 or 1 so that the resulting codeword has an even number of 1's. Thus, the check bit ensures that the number of 1's in a codeword is even, i.e. establishes *even parity.*

Example 10.

Let

$$H = \begin{bmatrix} 1 & 0 & 1 & 1 & 0 & 0 \\ 1 & 1 & 0 & 0 & 1 & 0 \\ 0 & 1 & 1 & 0 & 0 & 1 \end{bmatrix}$$

be a parity-check matrix for a binary (6,3)-code C. Which codeword carries the information $\mathbf{m} = (101)$? This, of course, depends on the basis chosen to generate C. Suppose we would like the information to appear in the first 3 components of the codeword. We must find check bits x_1, x_2, x_3 so that $\mathbf{c} = (1\ 0\ 1\ x_1\ x_2\ x_3)$ is a codeword. Thus,

$$H\mathbf{c}^T = 0$$

yielding

$$1 + 1 + x_1 = 0 \quad \text{or} \quad x_1 = 0$$
$$1 + x_2 = 0 \quad \text{or} \quad x_2 = 1$$
$$1 + x_3 = 0 \quad \text{or} \quad x_3 = 1$$

Hence $\mathbf{c} = (101011)$ is the desired codeword. In general here, the relation $H\mathbf{c}^T = 0$ determines the check bits x_i of the codeword $\mathbf{c} = (m_1m_2m_3x_1x_2x_3)$ corresponding to a given message $\mathbf{m} = (m_1m_2m_3)$ to be those x_i satisfying the system

$$\begin{array}{ccccccc} m_1 & + & & & m_3 & = & x_1 \\ m_1 & + & m_2 & & & = & x_2 \\ & & m_2 & + & m_3 & = & x_3 \end{array}$$

Each check bit x_i is a function of some subset of the message bits m_j, and conversely, row i of H specifies a subset of codeword bits on which to perform an even-parity check. Hence H is called the parity-check matrix.

As explained earlier, we could also determine the codeword $\mathbf{c}$ corresponding to the message $\mathbf{m} = (101)$ by finding the generator matrix G for C where $G = [I_3\, A]$. This is easily accomplished from the given H, writing $H = [-A^T\, I_3]$. Then

$$G = [I_3\, A] = \begin{bmatrix} 1 & 0 & 0 & 1 & 1 & 0 \\ 0 & 1 & 0 & 0 & 1 & 1 \\ 0 & 0 & 1 & 1 & 0 & 1 \end{bmatrix}.$$

Now $\mathbf{c} = \mathbf{m}G = (101011)$.

Much can be said about a linear code C by studying a parity-check matrix H for C. An important example is the relation between the distance of C and linear dependencies among the columns of H, as indicated by the following theorem.

Theorem 3.5. Let H be a parity-check matrix for an (n,k)-code C over F. Then every set of $s-1$ columns of H are linearly independent if and only if C has distance at least s.

Proof.

First, assume that every set of $s-1$ columns of H are linearly independent over F. Let $\mathbf{c} = (c_1c_2...c_n)$ be a non-zero codeword and let $\mathbf{h}_1, \mathbf{h}_2, \ldots, \mathbf{h}_n$ be the columns of H. Then since H is the parity-check matrix, $H\mathbf{c}^T = \mathbf{0}$. This matrix-vector product may be written in the form

$$Hc^T = \sum_{i=1}^{n} c_i \mathbf{h}_i = \mathbf{0}.$$

If $w(\mathbf{c}) \le s-1$, then we have a nontrivial linear combination of less than s columns of H which sums to $\mathbf{0}$. This is not possible since by hypothesis every set of $s-1$ or fewer columns of H are linearly independent. Therefore $w(\mathbf{c}) \ge s$, and since $\mathbf{c}$ is an arbitrary non-zero codeword and C is a linear code, it follows by Theorem 3.1 that $w(C) \ge s$, implying that C has distance at least s.

To prove the converse, assume that C has distance at least s. Suppose that some set of $t \le s-1$ columns of H are linearly dependent. Let these columns be $\mathbf{h}_{i_1}, \mathbf{h}_{i_2}, \ldots, \mathbf{h}_{i_t}$. Then there exist scalars $\lambda_{i_j} \in F$, not all zero, such that

$$\sum_{j=1}^{t} \lambda_{i_j} \mathbf{h}_{i_j} = 0.$$

Construct a vector $\mathbf{c}$ having λ_{i_j} in position i_j, $1 \le j \le t$, and 0's elsewhere. By construction, $\mathbf{c}$ is a non-zero vector in C since $H\mathbf{c}^T = \mathbf{0}$. But $w(\mathbf{c}) = t \le s-1$. This is a contradiction since by hypothesis every non-zero codeword in C has weight at least s. We conclude that no $s-1$ columns of H are linearly dependent. □

It follows easily from the theorem that a linear code C with parity-check matrix H has distance (exactly) d if and only if every set of $d-1$ columns of H are linearly independent, and some set of d columns are linearly dependent. Hence this theorem could be used to *determine* the distance of a linear code, given a parity-check matrix.

Theorem 3.5 is also a useful tool for *constructing* single-error correcting codes (i.e. codes with distance 3). To construct such a code, we need only construct a matrix H such that no 2 or fewer columns are linearly dependent. The only way a single column can be linearly dependent is if it is the zero column. Suppose two non-zero columns $\mathbf{h}_i$ and $\mathbf{h}_j$ are linearly dependent. Then there exist non-zero scalars $\lambda, \beta \in F$ such that

$$\lambda \mathbf{h}_i + \beta \mathbf{h}_j = \mathbf{0}.$$

This implies that

$$\mathbf{h}_i = -\lambda^{-1}\beta\mathbf{h}_j,$$

meaning $\mathbf{h}_i$ and $\mathbf{h}_j$ are scalar multiples of each other. Thus, if we construct H so that H contains no zero column and no two columns of H are scalar multiples, then H will be the parity-check matrix for a code having distance at least 3.

Example 11.

Over the field $F = Z_3$, consider the matrix

$$H = \begin{bmatrix} 1 & 0 & 0 & 1 & 2 \\ 0 & 2 & 0 & 0 & 1 \\ 0 & 0 & 1 & 1 & 0 \end{bmatrix}.$$

The columns of H are non-zero and no column is a scalar multiple of any other column. Hence, H is the parity-check matrix for a (5,2)-code C over F of distance at least 3. We can easily construct C if we know a generator matrix G for C. One can be found from H by constructing another parity-check matrix H' of the form $H' = [I_3\, A]$. Using row operations on H, we get

$$H' = \begin{bmatrix} 1 & 0 & 0 & 1 & 2 \\ 0 & 1 & 0 & 0 & 2 \\ 0 & 0 & 1 & 1 & 0 \end{bmatrix} = [I_3\, A], \qquad A = \begin{bmatrix} 1 & 2 \\ 0 & 2 \\ 1 & 0 \end{bmatrix}$$

$$G = [-A^T\, I_2] = \begin{bmatrix} 2\,0\,2\,1\,0 \\ 1\,1\,0\,0\,1 \end{bmatrix}.$$

To check we compute

$$HG^T = \begin{bmatrix} 1 & 0 & 0 & 1 & 2 \\ 0 & 2 & 0 & 0 & 1 \\ 0 & 0 & 1 & 1 & 0 \end{bmatrix} \begin{bmatrix} 2 & 1 \\ 0 & 1 \\ 2 & 0 \\ 1 & 0 \\ 0 & 1 \end{bmatrix} = \begin{bmatrix} 0 & 0 \\ 0 & 0 \\ 0 & 0 \end{bmatrix}.$$

Since $HG^T = 0$ and G contains 2 linearly independent rows, then G must generate the orthogonal complement to the space generated by H. Thus, G is the generator matrix for C. C will contain 9 codewords. We list the codewords and their weights.

Codeword	Weight
00000	0
20210	3
11001	3
10120	3
22002	3
01211	4
21121	5
12212	5
02122	4

It is easily verified that every pair of codewords has distance at least 3.

If the code C is over Z_2, then constructing H becomes even simpler, if we only require a distance 3 code. In this case, the columns of H need only be non-zero and distinct. This follows because in the binary case, the only way that one column $\mathbf{h}_i$ can be a scalar multiple of another column $\mathbf{h}_j$ is if $\mathbf{h}_i = \mathbf{h}_j$ or $\mathbf{h}_i = \mathbf{0}$, as 0 and 1 are the only scalars.

Example 12.

Suppose we wish to construct a parity-check matrix H for a single-error correcting (7,4)-code C over Z_2. Since C is a (7,4)-code, then $C^{\perp}$ is a (7,3)-code and H must be 3×7. If C is to have distance 3, then the columns of H must be distinct and non-zero. There are precisely 7 non-zero 3-tuples over Z_2 and so each must be a column of H. Therefore,

$$H = \begin{bmatrix} 1 & 0 & 0 & 1 & 0 & 1 & 1 \\ 0 & 1 & 0 & 1 & 1 & 0 & 1 \\ 0 & 0 & 1 & 0 & 1 & 1 & 1 \end{bmatrix}.$$

The order in which the columns are written is immaterial. Any other ordering gives an equivalent code. Again, we can easily construct a generator matrix for C. Since $H = [I_3\, A]$ where

$$A = \begin{bmatrix} 1 & 0 & 1 & 1 \\ 1 & 1 & 0 & 1 \\ 0 & 1 & 1 & 1 \end{bmatrix}$$

then

$$G = [-A^T\ I_4] = \begin{bmatrix} 1&1&0&1&0&0&0 \\ 0&1&1&0&1&0&0 \\ 1&0&1&0&0&1&0 \\ 1&1&1&0&0&0&1 \end{bmatrix}.$$

Suppose that we want to construct a parity-check matrix for a 2-error correcting code. The code must have distance at least 5, and hence by Theorem 3.5, no 4 columns of the parity-check matrix can be linearly dependent. Unfortunately, there is no simple condition guaranteeing this as in the case of a single-error correcting code and our task is more difficult. We will conclude this section by constructing a parity-check matrix for a 2-error correcting code. The construction is not an obvious one, but we shall see later that it is part of a very important general algebraic technique for constructing multiple-error-correcting codes.

Example 13.

Consider the matrix H given below, over the field $F = Z_7$. Note that 3 is a generating element in Z_7. That is, $Z_7 \backslash \{0\} = \{3^i : 0 \le i \le 5\}$.

$$H = \begin{bmatrix} 3^0 & 3^1 & 3^2 & 3^3 & 3^4 & 3^5 \\ 3^0 & 3^2 & 3^4 & 3^6 & 3^8 & 3^{10} \\ 3^0 & 3^3 & 3^6 & 3^9 & 3^{12} & 3^{15} \\ 3^0 & 3^4 & 3^8 & 3^{12} & 3^{16} & 3^{20} \end{bmatrix} = \begin{bmatrix} 1&3&2&6&4&5 \\ 1&2&4&1&2&4 \\ 1&6&1&6&1&6 \\ 1&4&2&1&4&2 \end{bmatrix}.$$

We claim that H is a parity-check matrix for a (6,2)-code C over Z_7 and that C has distance 5. We must prove that no 4 columns of H are linearly dependent. We could do this by checking all $\binom{6}{4} = 15$ possibilities. Instead, we look at the general situation. Take any 4 columns of H and form a matrix D.

$$D = \begin{bmatrix} 3^i & 3^j & 3^k & 3^l \\ 3^{2i} & 3^{2j} & 3^{2k} & 3^{2l} \\ 3^{3i} & 3^{3j} & 3^{3k} & 3^{3l} \\ 3^{4i} & 3^{4j} & 3^{4k} & 3^{4l} \end{bmatrix} \quad \text{where } i < j < k < l.$$

If the columns of D are linearly independent, then the *determinant* of D must be non-zero and conversely so. We compute $\det D$. Recall from linear algebra that if a matrix D' is derived from D by multiplying a column by the scalar s, then $\det D' = s \det D$. Using this, we get

$$\det D = 3^i 3^j 3^k 3^l \begin{vmatrix} 1 & 1 & 1 & 1 \\ 3^i & 3^j & 3^k & 3^l \\ 3^{2i} & 3^{2j} & 3^{2k} & 3^{2l} \\ 3^{3i} & 3^{3j} & 3^{3k} & 3^{3l} \end{vmatrix}$$

Now we multiply row 3 by 3^i and subtract this from row 4; we multiply row 2 by 3^i and subtract from row 3; and so on. Recall that adding scalar multiples of one row to another does not affect the determinant. (For completeness, although we do not use it here, recall that the third elementary row operation, the exchange of two rows, causes the determinant to be negated.) We get

$$\det D = 3^{i+j+k+l} \begin{vmatrix} 1 & 1 & 1 & 1 \\ 0 & 3^j-3^i & 3^k-3^i & 3^l-3^i \\ 0 & 3^j(3^j-3^i) & 3^k(3^k-3^i) & 3^l(3^l-3^i) \\ 0 & 3^{2j}(3^j-3^i) & 3^{2k}(3^k-3^i) & 3^{2l}(3^l-3^i) \end{vmatrix}$$

$$= 3^{i+j+k+l}(3^j-3^i)(3^k-3^i)(3^l-3^i) \begin{vmatrix} 1 & 1 & 1 \\ 3^j & 3^k & 3^l \\ 3^{2j} & 3^{2k} & 3^{2l} \end{vmatrix}$$

$$= 3^{i+j+k+l}(3^j-3^i)(3^k-3^i)(3^l-3^i)\cdot(3^k-3^j)(3^l-3^j)\cdot(3^l-3^k)$$

$$= 3^{i+j+k+l} \prod_{x>y} (3^x-3^y)$$

where the product is over $x,y \in \{i,j,k,l\}$, $x > y$. Since $3^x - 3^y \neq 0$ for $x \neq y$, it follows that $\det D \neq 0$ and no 4 columns of H are linearly dependent. Therefore, H is the parity-check matrix for a (6,2)-code over Z_7 with distance at least 5. The code C will have 7^2 code-words.

The technique of the preceding example will be considered in greater detail in Chapter 6 when we discuss *BCH-codes*.

3.4 Hamming Codes and Perfect Codes

In this section, we will construct a very interesting class of single-error correcting codes. These codes are known as *Hamming codes*. They will be defined in terms of a parity-check matrix.

Definition. A *Hamming code* of order r over $GF(q)$ is an (n,k)-code where $n=(q^r-1)/(q-1)$ and $k=n-r$, with parity-check matrix H_r an $r\times n$ matrix such that the columns of H_r are non-zero and no two columns are scalar multiples of each other.

By Theorem 3.5, it follows immediately that the Hamming codes have distance $d=3$ and are hence single-error correcting. Note that q^r-1 is the number of non-zero r-tuples that exist, and $q-1$ is the number of non-zero scalars. Hence if any additional column was appended to H_r, this column would be a scalar multiple of one already present, and this would be undesirable if we want a single-error correcting code (by Theorem 3.5). In this sense then, the codewords have maximum length, and hence carry a maximum number of information symbols, for a given value of r, giving the code a maximum number of codewords for such r.

Example 14.

The Hamming code of order $r=3$ over $F=GF(2)$ is defined by the parity check matrix

$$H_3=\begin{bmatrix}1&0&0&1&0&1&1\\0&1&0&1&1&0&1\\0&0&1&0&1&1&1\end{bmatrix}.$$

This is a (7,4)-code with distance 3. Re-ordering the columns of H_3 would define an equivalent Hamming code.

Example 15.

Take $r=3$, $F=GF(3)$. Then H_r defines the (13,10) Hamming code of order 3 over $GF(3)$, where

$$H_3=\begin{bmatrix}1&0&0&1&0&1&1&2&0&1&2&1&1\\0&1&0&1&1&0&1&1&2&0&1&2&1\\0&0&1&0&1&1&1&0&1&2&1&1&2\end{bmatrix}.$$

The Hamming codes are interesting because every n-tuple is in a sphere of radius one about some codeword. They are an example of a *perfect code.*

Definition. A *perfect code* is an e-error-correcting $[n,M]$-code over an alphabet A such that every n-tuple over A is in the sphere of radius e about some codeword.

In a perfect code of block length n, the spheres of radius e about the codewords are not only disjoint (since the code is e-error-correcting) but exhaust the entire space of n-tuples. We see that it is easy to prove that the Hamming codes are perfect.

Theorem 3.6. The Hamming code C of order r over $GF(q)$ is a perfect code.

Proof.

As noted earlier, the Hamming codes have distance 3 so that $e = 1$. Consider a codeword $\mathbf{c} \in C$. We first determine the number of vectors in the sphere of radius one about $\mathbf{c}$. Vectors of distance one from $\mathbf{c}$ are produced by choosing one of n coordinate positions in $\mathbf{c}$, and altering that component to any of $q-1$ values. Hence including $\mathbf{c}$ itself, the sphere of radius one about $\mathbf{c}$ contains $1 + n(q-1)$ vectors. Since spheres about distinct codewords are disjoint, and since there are q^k codewords, where $k = n-r$, the total number of vectors contained in all spheres is

$$[1 + n(q-1)]q^k = [1 + (q^r-1)]q^{n-r} = q^n.$$

But this is precisely the total number of vectors in the entire space of n-tuples over $GF(q)$. Hence every n-tuple is in some sphere and the code is perfect. □

Examining this proof, it follows from the definition that an e-error-correcting $[n,M]$-code is perfect if and only if it meets the *sphere-packing bound* (*Hamming bound*)

$$q^n \geq M \sum_{i=0}^{e} \binom{n}{i}(q-1)^i$$

with equality.

We conclude this section by mentioning that the parameters for perfect codes over alphabets having a prime power number of elements have been completely determined. The proof is beyond the scope of this book.

3.5 Decoding Single-Error Correcting Codes

Linear codes such as the Hamming codes are by far the simplest types of codes to decode. In this section we outline one decoding procedure for linear codes capable of correcting a single error per codeword. We begin by introducing the concept of an *error vector,* providing an algebraic way to manipulate errors.

Example 16.

Suppose $\mathbf{c} = (0\ 1\ 0\ 1\ 1\ 1)$ is a binary codeword which is transmitted and

$$\mathbf{r} = (1\ 0\ 0\ 1\ 1\ 1)$$

is received. We can represent $\mathbf{r}$ as

$$\mathbf{r} = \mathbf{c} + \mathbf{e}$$

where $\mathbf{e} = (1\ 1\ 0\ 0\ 0\ 0)$. We call $\mathbf{e}$ the error vector introduced by the channel.

Let H be a parity-check matrix for a linear code C. Suppose $w(\mathbf{e}) \leq 1$ for any error vector introduced by the channel, so that a single-error-correcting code can correct all error patterns introduced. Suppose $\mathbf{c} \in C$ is a transmitted codeword and $\mathbf{r}$ is received. Write $\mathbf{r} = \mathbf{c} + \mathbf{e}$. Compute

$$H\mathbf{r}^T = H(\mathbf{c}+\mathbf{e})^T = H\mathbf{c}^T + H\mathbf{e}^T = H\mathbf{e}^T$$

since $H\mathbf{c}^T = 0$. If $\mathbf{e} = 0$, then $H\mathbf{e}^T = 0$. Hence if $H\mathbf{r}^T = 0$, we accept $\mathbf{r}$ as the transmitted codeword. If $\mathbf{e} \neq 0$, then $\mathbf{e}$ contains one non-zero coordinate. Suppose this is the ith coordinate and its value is α. Then $H\mathbf{e}^T = \alpha\mathbf{h}_i$, where $\mathbf{h}_i$ is the ith column of H. This observation gives us a decoding procedure.

Decoding Single-Error Linear Codes.

Let H be the parity-check matrix, and let $\mathbf{r}$ be the received vector.

(1) Compute $H\mathbf{r}^T$.

(2) If $H\mathbf{r}^T = 0$, then accept $\mathbf{r}$ as the transmitted codeword.

(3) If $H\mathbf{r}^T = \mathbf{s}^T \neq 0$, then compare $\mathbf{s}^T$ with the columns of H.

(4) If there is some i such that $\mathbf{s}^T = \alpha\,\mathbf{h}_i$, then $\mathbf{e}$ is the n-tuple with α in position i and 0's elsewhere; correct $\mathbf{r}$ to $\mathbf{c} = \mathbf{r} - \mathbf{e}$.

(5) Otherwise, more than one error has occurred.

Example 17.

Consider the single-error-correcting code C over Z_2 having parity-check matrix

$$H = \begin{bmatrix} 1 & 0 & 0 & 1 & 1 & 0 \\ 0 & 1 & 0 & 1 & 1 & 1 \\ 0 & 0 & 1 & 0 & 1 & 1 \end{bmatrix}.$$

Suppose the codeword $\mathbf{c} = (1\ 1\ 0\ 1\ 0\ 0)$ is transmitted and the channel introduces the error pattern $\mathbf{e} = (0\ 0\ 0\ 0\ 1\ 0)$ so that the receiver gets $\mathbf{r} = (1\ 1\ 0\ 1\ 1\ 0)$. Compute

$$H\mathbf{r}^T = \begin{bmatrix} 1 \\ 1 \\ 1 \end{bmatrix}.$$

Since this is the 5th column of H we conclude that the error occurred in position 5 and the error pattern is $\mathbf{e} = (0\ 0\ 0\ 0\ 1\ 0)$. Note that over Z_2,

$$\mathbf{c} = \mathbf{r} - \mathbf{e} = \mathbf{r} + \mathbf{e}.$$

Example 18.

Consider the single-error-correcting code C over Z_5 having parity-check matrix

$$H = \begin{bmatrix} 1 & 0 & 0 & 2 & 4 & 1 & 0 \\ 0 & 2 & 0 & 1 & 0 & 2 & 2 \\ 0 & 0 & 3 & 1 & 4 & 1 & 2 \end{bmatrix}.$$

C consists of 5^4 codewords. Suppose the codeword $\mathbf{c} = (4\,4\,3\,0\,0\,1\,0)$ is transmitted and the channel introduces the error pattern $\mathbf{e} = (0\,0\,0\,3\,0\,0\,0)$. The receiver gets $\mathbf{r} = \mathbf{c} + \mathbf{e} = (4\,4\,3\,3\,0\,1\,0)$. Compute

$$H\mathbf{r}^T = \begin{bmatrix} 1 \\ 3 \\ 3 \end{bmatrix} = \mathbf{s}^T.$$

$\mathbf{s}^T$ must be a scalar multiple of some column of H. By examining each column of H and taking scalar multiples we see that $\mathbf{s}^T$ is 3 times column 4. This gives us the error pattern $\mathbf{e}$. The codeword is $\mathbf{c} = \mathbf{r} - \mathbf{e}$.

There is an interesting extension of the above technique for the binary Hamming codes. Suppose H is a parity-check matrix for such a code. Since the columns of H are all of the non-zero binary r-tuples, order the columns of H so that column i is the binary representation of the integer i. Now in the preceding algorithm we need not compare $\mathbf{s}^T$ with each column of H. The integer which $\mathbf{s}^T$ represents specifies the error location.

Example 19.

Consider the binary Hamming code with parity-check matrix

$$H = \begin{bmatrix} 1 & 0 & 1 & 0 & 1 & 0 & 1 \\ 0 & 1 & 1 & 0 & 0 & 1 & 1 \\ 0 & 0 & 0 & 1 & 1 & 1 & 1 \end{bmatrix}.$$

Notice that, for example, column 6 is $(0\,1\,1)^T$, representing the integer $0{\cdot}2^0 + 1{\cdot}2^1 + 1{\cdot}2^2 = 6$. Suppose that $\mathbf{c} = (0\,1\,1\,1\,1\,0\,0)$ is a transmitted codeword and the error $\mathbf{e} = (0\,0\,0\,0\,0\,1\,0)$ is introduced. The received vector is then $\mathbf{r} = (0\,1\,1\,1\,1\,1\,0)$ and $H\mathbf{r}^T = (0\,1\,1)^T$. The binary interpretation of this vector reveals bit 6 to be in error.

3.6 Standard Array Decoding and Syndrome Decoding

The decoding scheme of the previous section is efficient for single-error-correcting codes, but fails for linear codes that can correct more than one error. In this section we describe a general technique for decoding any linear code.

Before we introduce *standard array decoding*, we recall some basic concepts from group theory. A *group* is a set of elements together with an operation, such that the set is closed under the operation, the operation is associative, an identity element exists, and all elements have inverses. Formally, we have the following definition.

Definition. A *group* $(G,*)$ consists of a set of elements together with an operation * such that

(i) if $a,b \in G$ then $a*b \in G$

(ii) if $a,b,c \in G$ then $(a*b)*c = a*(b*c)$

(iii) there exists an element $e \in G$ such that $e*a = a*e = a$ for all $a \in G$

(iv) for each $a \in G$ there exists an element $a^{-1} \in G$ such that $a*a^{-1} = a^{-1}*a = e$.

If, in addition to these properties, $a*b = b*a$ for all $a,b \in G$ (the operation is commutative), then G is called an *abelian group*. We will be interested in finite groups, i.e. groups which contain only a finite number of elements.

Example 20.

Take G to be the vectors in $V_n(F)$ and * to be vector addition. This system satisfies the axioms of a group. For example for n=2 and F=$GF(2)$,

$$G = \{(00), (10), (01), (11)\}$$

is a group under vector addition, with

$$e = (0,0)$$

$$(00)^{-1} = (00), \ (10)^{-1} = (10), \ (01)^{-1} = (01), \ (11)^{-1} = (11).$$

A note regarding notation is in order. We often distinguish a group as being either an *additive* group or a *multiplicative* group, if one seems more natural than the other (although from a group-theoretical point of view, the two are equivalent). For an additive group, we call the group operation *addition*, and use + instead of *. The inverse of an element is denoted $-a$ instead of a^{-1}, and the identity element is denoted 0. When viewed as a multiplicative group, the operation is referred to as *multiplication*, and we frequently write $a \cdot b$ (or simply ab) in place of $a*b$. We denote the inverse of an element by either a^{-1} or $1/a$, and use 1 to denote the identity. As examples, for an additive group, consider the integers mod n (under addition), and for a multiplicative group, consider the non-zero integers mod p, p a prime (under multiplication). Example 20 above may appear more familiar if the inverses are denoted in additive notation.

A *subgroup* H of a finite group G is a nonempty subset of G which is closed under the group operation. In other words, H together with the operation in G is itself a group. If H is a subgroup of G we denote this $H \subseteq G$. We can use such a subgroup to define a relation between elements of the group as follows.

Definition. Let H be a subgroup of a group G with $g_1, g_2 \in G$. We say g_1 is *congruent* to g_2 *modulo H*, and denote this $g_1 \equiv g_2 \pmod{H}$, if and only if $g_1 g_2^{-1} \in H$.

In the additive notation, $g_1 \equiv g_2 \pmod{H}$ if $g_1 - g_2 \in H$. Just as in the case of congruence modulo a polynomial in §2.1, congruence modulo a subgroup is an equivalence relation which partitions the elements into disjoint equivalence classes (exercise 51). For $H \subseteq G$ the equivalence classes for congruence modulo H are called *right cosets* of H. The right coset of H which contains the element $g \in G$ is defined to be

$$[g] = \{x \in G\colon x \equiv g \pmod{H}\}.$$

Since $x \equiv g \pmod{H}$ means $xg^{-1} \in H$, there is some element $h \in H$ such that $xg^{-1} = h$. Hence $x = hg$, and we see that an equivalent definition is

$$[g] = \{hg\colon h \in H\} = Hg,$$

or in additive notation,

$$[g] = \{h+g\colon h \in H\} = H+g.$$

H itself is a right coset (it is the coset He), and if G is an abelian group (as it will be for all of our work), the right cosets are simply called *cosets*. With this, we have that $g_1 \equiv g_2 \pmod{H}$ if and only if g_1 and g_2 are in the same coset of H.

Example 21.

Consider the group

$$G = \{(00), (10), (01), (11)\}$$

with group operation vector addition mod 2, and the subgroup $H = \{(00), (10)\}$. The cosets of H are $H+(00) = H$ and

$$H+(01) = \{(01), (11)\} = H+(11).$$

Since (01) and (11) are in the same coset of H, they are congruent modulo H. Working straight from the definition, $(01) \equiv (11) \pmod{H}$ since $(01) - (11) = (10) \in H$.

The *order* of a group G is the number of elements in G. The following theorem is fundamental in the study of finite groups.

Theorem 3.7 *(Lagrange's Theorem).* **If G is a group of order n, and H is a subgroup of order m, then m divides n.**

Proof.

Let Hg_1 and Hg_2 be two cosets of H. Let $H = \{h_1, h_2, \ldots, h_m\}$. We define the mapping $f: Hg_1 \to Hg_2$ by $f(h_i g_1) = h_i g_2$, $1 \le i \le m$. If f maps two elements of Hg_1 to the same element of Hg_2, that is, if $f(h_i g_1) = f(h_j g_1)$, then $h_i g_2 = h_j g_2$. Hence $h_i = h_j$, implying $h_i g_1 = h_j g_1$, and the two elements are in fact the same. So f maps distinct elements of Hg_1 to distinct elements of Hg_2. Also, it follows directly from the definition of f that each element of Hg_2 gets mapped onto. Thus, f is a one-to-one mapping and Hg_1 and Hg_2 have the same cardinality. Hence, all cosets of H have cardinality m. Since congruence modulo H is an equivalence relation, the cosets of H partition G. It follows that

$$n = m \times (\text{the number of cosets of } H)$$

and that m divides n. □

We are now in a position to construct a *standard array* for an (n,k)-code C over F. We can think of C as a subgroup of order q^k in the group $V_n(F)$. (C is, of course, also a subspace.) C has $t = q^n/q^k = q^{n-k}$ cosets. Denote these cosets by $C_0, C_1, \ldots, C_{t-1}$ where $C_0 = C$. For each C_i, $0 \le i \le t-1$, let $\mathbf{l}_i$ be a vector of minimum weight in C_i. The $\mathbf{l}_i$ are called *coset leaders*. Note that $\mathbf{l}_0 = \mathbf{0}$. If $C_0 = \{\mathbf{l}_0, \mathbf{c}_1, \mathbf{c}_2, \ldots, \mathbf{c}_{q^k-1}\}$ are the codewords of C, then form a $q^{n-k} \times q^k$ array S where row i+1, column j+1 contains the entry $\mathbf{l}_i + \mathbf{c}_j$. Hence the entries in row i+1 are the elements in coset C_i, and the entries in column 1 are the coset leaders. Note that S contains all vectors in $V_n(F)$, and C_i is the coset $C+\mathbf{l}_i$. S is called a standard array for the code C.

Example 22.

Table 1 gives a standard array for the binary (5,2)-code with generator matrix

$$G = \begin{bmatrix} 1 & 0 & 1 & 0 & 1 \\ 0 & 1 & 1 & 1 & 0 \end{bmatrix}.$$

Coset leaders			
00000	10101	01110	11011
00001	10100	01111	11010
00010	10111	01100	11001
00100	10001	01010	11111
01000	11101	00110	10011
10000	00101	11110	01011
11000	01101	10110	00011
10010	00111	11100	01001

Table 1. A standard array for the (5,2)-code of example 22.

Example 23.

A standard array for the binary (6,3)-code with generator matrix G and parity-check matrix H, where

$$G=\begin{bmatrix}1&1&0&1&0&0\\0&1&1&0&1&0\\1&0&1&0&0&1\end{bmatrix},\qquad H=\begin{bmatrix}1&0&0&1&0&1\\0&1&0&1&1&0\\0&0&1&0&1&1\end{bmatrix}$$

is given in Table 2. The codewords are the vectors in the first row of the table. Note that this code has distance 3.

Coset leaders							
000000	110100	011010	101001	101110	110011	011101	000111
000001	110101	011011	101000	101111	110010	011100	000110
000010	110110	011000	101011	101100	110001	011111	000101
000100	110000	011110	101101	101010	110111	011001	000011
001000	111100	010010	100001	100110	111011	010101	001111
010000	100100	001010	111001	111110	100011	001101	010111
100000	010100	111010	001001	001110	010011	111101	100111
001100	111000	010110	100101	100010	111111	010001	001011

Table 2. A standard array for the (6,3)-code of example 23.

An important property of standard arrays, providing the basis for a decoding algorithm, is given by the following result.

Lemma 3.8. Let C be an (n,k)-code over $GF(q)$ with codewords $\mathbf{c}_j$, $0 \le j \le q^k-1$. Let S be a standard array with coset leaders $\mathbf{l}_i$, $0 \le i \le q^{n-k}-1$, and $(i+1,j+1)$-entry $\mathbf{l}_i+\mathbf{c}_j$. Then for each i and j, for all h, $0 \le h \le q^k-1$, we have

$$d(\mathbf{l}_i+\mathbf{c}_j, \mathbf{c}_j) \le d(\mathbf{l}_i+\mathbf{c}_j, \mathbf{c}_h).$$

Proof.

Expressing the distance between vectors as the weight of their difference vector, we have

$$d(\mathbf{l}_i+\mathbf{c}_j, \mathbf{c}_j) = w(\mathbf{l}_i)$$

$$d(\mathbf{l}_i+\mathbf{c}_j, \mathbf{c}_h) = w(\mathbf{l}_i+\mathbf{c}_j-\mathbf{c}_h).$$

Since the coset $C_i = C+\mathbf{l}_i$ has coset leader $\mathbf{l}_i$, and $\mathbf{l}_i+(\mathbf{c}_j-\mathbf{c}_h) \in C_i$,

$$w(\mathbf{l}_i) \le w(\mathbf{l}_i+\mathbf{c}_j-\mathbf{c}_h).$$

The result follows. □

In other words, if $\mathbf{r} = \mathbf{l}_i+\mathbf{c}_j$ is a received vector, then the codeword $\mathbf{c}_j$ is as close to $\mathbf{r}$ as any other codeword. We observe from the proof of this theorem that if $\mathbf{l}_i$ is the ***unique*** vector of least weight in the coset C_i, then $\mathbf{c}_j$ is the closest codeword to $\mathbf{l}_i + \mathbf{c}_j$, so that by nearest neighbour decoding, $\mathbf{r}$ should be decoded to $\mathbf{c}_j$. (If $\mathbf{l}_i$ is not unique, then $\mathbf{c}_j$ is still as close as any other codeword $\mathbf{c}_h$.)

Of interest now is knowledge as to when a coset leader is the unique vector of minimum weight in its coset. This brings us to the following important property of a standard array.

Theorem 3.9. Let C be a linear code with distance d. If $\mathbf{x}$ is a vector such that

$$w(\mathbf{x}) \le \lfloor \frac{d-1}{2} \rfloor,$$

then $\mathbf{x}$ is a ***unique*** element of minimum weight in its coset of C and hence is always a coset leader in a standard array for C.

Proof.

Suppose $w(\mathbf{x}) \le \lfloor (d-1)/2 \rfloor$ and $\mathbf{x}$ is not a unique vector of minimum weight in its coset C_i. Then there exists some vector $\mathbf{y} \in C_i$ such that $w(\mathbf{y}) \le w(\mathbf{x})$. Now since $\mathbf{x}$ and $\mathbf{y}$ are in the same coset C_i of C, $\mathbf{x} \equiv \mathbf{y} \pmod{C}$ and $\mathbf{x}-\mathbf{y} \in C$. But

$$w(\mathbf{x}-\mathbf{y}) \le w(\mathbf{x}) + w(\mathbf{y}) \le \lfloor \frac{d-1}{2} \rfloor + \lfloor \frac{d-1}{2} \rfloor \le d-1.$$

This contradicts the distance d of the code, unless $\mathbf{x}-\mathbf{y}$ is the zero vector, i.e. $\mathbf{x} = \mathbf{y}$. The result follows. □

The preceding properties of the standard array suggest the following decoding scheme.

Standard Array Decoding for Linear Codes.

Precomputation: Construct a standard array S.

Let $\mathbf{r}$ be a received vector.

(1) Find $\mathbf{r}$ in the standard array S.

(2) Correct $\mathbf{r}$ to the codeword at the top of its column.

Since $\mathbf{r}$ is in some coset C_i, then $\mathbf{r} = \mathbf{l}_i + \mathbf{c}_j$ for some codeword $\mathbf{c}_j$. Correcting to the top of the column can be accomplished by subtracting the coset leader from the received word. That is, $\mathbf{c}_j = \mathbf{r} - \mathbf{l}_i$. We notice that any error pattern which corresponds to a coset leader will be corrected. If C is a distance d code, then Theorem 3.9 says that every error pattern of weight less than or equal to $e = \lfloor (d-1)/2 \rfloor$ appears as a coset leader in S. Thus, S will correct any e or fewer errors. The decoding scheme outlined above will also correct those errors of weight $e+1$ whose patterns appear as coset leaders.

Example 24.

Consider the (6,3)-code C and standard array given in Table 2.

(i) Suppose $\mathbf{r} = (1\,0\,0\,1\,1\,0)$ is received. $\mathbf{r}$ occurs in row 5, column 5. We correct $\mathbf{r}$ to the codeword $(1\,0\,1\,1\,1\,0)$ at the top of column 5. The error pattern which occurred is the coset leader $(0\,0\,1\,0\,0\,0)$, assuming at most one error occurred.

(ii) Suppose $\mathbf{c} = (0\,1\,1\,0\,1\,0)$ is the transmitted codeword and $\mathbf{r} = (0\,1\,1\,0\,1\,1)$ is received. $\mathbf{r}$ occurs in row 2 column 3. Row 2 has coset leader $\mathbf{l} = (0\,0\,0\,0\,0\,1)$, which is the error pattern. $\mathbf{r}$ is corrected to $\mathbf{c} = \mathbf{r} - \mathbf{l}$.

(iii) Suppose $\mathbf{c} = (0\ 1\ 1\ 0\ 1\ 0)$ is the codeword transmitted and $\mathbf{r} = (1\ 1\ 1\ 0\ 1\ 1)$ is received. $\mathbf{r}$ occurs in row 5, column 6 and so following the above scheme, $\mathbf{r}$ is decoded to the codeword $(1\ 1\ 0\ 0\ 1\ 1)$. The decoding in incorrect in this case. The error pattern $(1\ 0\ 0\ 0\ 0\ 1)$ has weight 2, but the code, having distance 3, is only 1-error-correcting in general. It cannot correct every error pattern of weight 2, and since this error pattern is not a coset leader, it is not correctable.

(iv) Suppose $\mathbf{c} = (1\ 1\ 0\ 0\ 1\ 1)$ is transmitted and the error $(0\ 0\ 1\ 1\ 0\ 0)$ occurs, so that $\mathbf{r} = (1\ 1\ 1\ 1\ 1\ 1)$ is received. We locate $\mathbf{r}$ in row 6, column 6 and correct it to $\mathbf{c}$. Since in this case the error pattern, although of weight 2, is a coset leader, it is still correctable.

We see that the standard array of Table 2 corrects all error patterns of weight 1 and one pattern of weight 2.

The decoding scheme given above, although very simple, is quite inefficient. Storing the standard array is impractical for large codes, and locating a received vector within the table is a nontrivial task. We can improve the situation with the aid of a few observations.

Definition. Let H be a parity-check matrix for an (n,k)-code over F. For $\mathbf{x} \in V_n(F)$, the *syndrome* $\mathbf{s}$ of $\mathbf{x}$ is defined to be $\mathbf{s} = H\mathbf{x}^T$.

The decoding algorithm of §3.5 made use of the syndrome. We shall use it later in this section in a more general decoding algorithm.

Example 25.

$$H = \begin{bmatrix} 1 & 0 & 0 & 1 & 0 \\ 0 & 1 & 0 & 0 & 1 \\ 0 & 0 & 1 & 1 & 1 \end{bmatrix}$$

is the parity-check matrix for a (5,2)-code over Z_2. The syndrome of $\mathbf{x} = (0\ 0\ 1\ 1\ 0)$ is

$$H\mathbf{x}^T = \begin{bmatrix} 1 & 0 & 0 & 1 & 0 \\ 0 & 1 & 0 & 0 & 1 \\ 0 & 0 & 1 & 1 & 1 \end{bmatrix} \begin{bmatrix} 0 \\ 0 \\ 1 \\ 1 \\ 0 \end{bmatrix} = \begin{bmatrix} 1 \\ 0 \\ 0 \end{bmatrix}.$$

Theorem 3.10. Let H be a parity-check matrix for a linear code C. Then two vectors **x** and **y** are in the same coset of C if and only if they have the same syndrome (i.e. $H\mathbf{x}^T = H\mathbf{y}^T$).

Proof.

If **x** and **y** are in the same coset of C, then $\mathbf{x} = \mathbf{l} + \mathbf{c}_i$ and $\mathbf{y} = \mathbf{l} + \mathbf{c}_j$ for some codewords $\mathbf{c}_i$, $\mathbf{c}_j$ and some coset leader **l**. Now

$$H\mathbf{x}^T = H(\mathbf{l}+\mathbf{c}_i)^T = H\mathbf{l}^T + H\mathbf{c}_i^T = H\mathbf{l}^T$$

since $H\mathbf{c}_i^T = 0$. Similarly, $H\mathbf{y}^T = H\mathbf{l}^T$. Hence $H\mathbf{x}^T = H\mathbf{y}^T$.

Conversely, suppose $H\mathbf{x}^T = H\mathbf{y}^T$. Then $H\mathbf{x}^T - H\mathbf{y}^T = 0$ and $H(\mathbf{x}-\mathbf{y})^T = 0$ implying $\mathbf{x} - \mathbf{y}$ is a codeword. Hence $\mathbf{x} \equiv \mathbf{y} \pmod{C}$, and **x** and **y** are in the same coset. □

The word *syndrome* is from the Greek meaning *a running together* or *combination*. It is an appropriate name for the common vector associated with the vectors of a coset.

We can use syndromes to simplify the standard array decoding scheme given earlier, by establishing at the outset the one-to-one correspondence between coset leaders and syndromes. The revised algorithm is as follows.

Syndrome Decoding for Linear Codes.

Precomputation: Set up the one-to-one correspondence between coset leaders and syndromes.

Let **r** be a received vector, and let H be the parity-check matrix.

(1) Compute the syndrome $\mathbf{s} = H\mathbf{r}^T$ of **r**.

(2) Find the coset leader **l** associated with **s**.

(3) Correct **r** to $\mathbf{r} - \mathbf{l}$.

Example 26.

For the (6,3)-code of Table 2, we establish the following correspondence.

Coset Leader	Syndrome
000000	000
000001	101
000010	011
000100	110
001000	001
010000	010
100000	100
001100	111

By Theorem 3.9, we know that all weight-1 vectors must be coset leaders, and hence the only leader not immediately obvious is the last, (001100). One might discover this leader by trying all weight-2 vectors until one with syndrome distinct from the first 7 syndromes is found. Now suppose $\mathbf{r} = (1\ 0\ 0\ 0\ 1\ 1)$ is received. Then $H\mathbf{r}^T = (0\ 1\ 0)^T$. Since $(0\ 1\ 0)$ is associated with the coset leader $\mathbf{l} = (0\ 1\ 0\ 0\ 0\ 0)$, we correct $\mathbf{r}$ to

$$\mathbf{r} - \mathbf{l} = (1\ 0\ 0\ 0\ 1\ 1) - (0\ 1\ 0\ 0\ 0\ 0) = (1\ 1\ 0\ 0\ 1\ 1).$$

If C is a binary (70,50)-code, then C has 2^{50} codewords. The number of cosets is $2^{70}/2^{50} = 2^{20}$. One way to implement the revised algorithm would be to store 70 bits for each coset leader and 20 bits for each syndrome (see also exercise 31). This would require

$$70 \times 2^{20} + 20 \times 2^{20} = 90 \times 2^{20}$$

bits, about 11 *megabytes*. We can cut this storage down using the step-by-step decoding method described in the next section.

3.7 Step-by-Step Decoding†

In this section we describe another decoding technique, called *step-by-step decoding*, which is based on standard array decoding. For simplicity, we describe the algorithm here in the binary case; it can be generalized (exercise 53).

With step-by-step decoding we need only keep a one-to-one correspondence between syndromes and *weights* of coset leaders. For the standard array of Table 2 of §3.6, we set up the following correspondence.

† This section may be omitted without loss of continuity.

Syndrome	Weight of coset leader
000	0
101	1
011	1
110	1
001	1
010	1
100	1
111	2

Then received vectors can be decoded using the following algorithm. Let $\mathbf{e}_i$ denote the binary n-tuple whose only non-zero component is component i.

Step-by-step Decoding Algorithm for Linear Codes.

Precomputation: Set up the one-to-one correspondence between syndromes and weights of the associated coset leaders. (This is determined by the chosen parity check matrix.)

Let H be the parity-check matrix, and $\mathbf{r} = (r_1 r_2 ... r_n)$ be the received word.

(1) Set $i = 1$

(2) Compute $H\mathbf{r}^T$ and the weight w of the corresponding coset leader.

(3) If $w = 0$, stop with $\mathbf{r}$ as the transmitted codeword.

(4) If $H(\mathbf{r}+\mathbf{e}_i)^T$ has smaller associated weight than $H\mathbf{r}^T$, set $\mathbf{r} = \mathbf{r} + \mathbf{e}_i$.

(5) Set $i = i+1$ and go to (2).

The algorithm essentially processes $\mathbf{r}$ flipping one bit at a time, and determines whether this action moves the vector into a coset with a lighter coset leader. If so, the bit is left in the altered condition, and otherwise the bit is restored. Throughout the algorithm, the associated weight is non-increasing. When the vector is moved into the coset with leader of weight 0, the coset is the code itself, and the vector has been decoded to a codeword.

Example 27.

Consider the (6,3)-code of Table 2 with parity-check matrix H. The one-to-one correspondence between syndromes and weights of associated coset leaders is given above. Suppose the received vector is

$$\mathbf{r} = (1\,1\,1\,0\,0\,0).$$

$H\mathbf{r}^T = (1\ 1\ 1)^T$ with associated weight 2. Now $\mathbf{r} + \mathbf{e}_1 = (0\,1\,1\,0\,0\,0)$, and $H(\mathbf{r}+\mathbf{e}_1)^T = (0\ 1\ 1)^T$ which has associated weight 1. Since the associated weight has decreased from 2 to 1, update **r** to

$$\mathbf{r} = \mathbf{r} + \mathbf{e}_1 = (0\,1\,1\,0\,0\,0).$$

Moving on to component 2 of **r**,

$$\mathbf{r} + \mathbf{e}_2 = (0\,0\,1\,0\,0\,0).$$

$H(\mathbf{r}+\mathbf{e}_2)^T = (0\ 0\ 1)^T$ with associated weight 1. Since the associated weight has not decreased, **r** remains as it was, and we move on to component 3.

$$\mathbf{r} + \mathbf{e}_3 = (0\,1\,0\,0\,0\,0)$$

which has syndrome (0 1 0) and associated weight 1, which is not smaller than the current value. Moving on,

$$\mathbf{r} + \mathbf{e}_4 = (0\,1\,1\,1\,0\,0).$$

$H(\mathbf{r}+\mathbf{e}_4)^T = (1\ 0\ 1)$ with associated weight 1. We move on.

$$\mathbf{r} + \mathbf{e}_5 = (0\,1\,1\,0\,1\,0).$$

$H(\mathbf{r}+\mathbf{e}_5)^T = (0\ 0\ 0)^T$ with associated weight 0. We set $\mathbf{r} = \mathbf{r} + \mathbf{e}_5 = (0\ 1\ 1\ 0\ 1\ 0)$, and this is the transmitted word.

To understand how the step-by-step decoding algorithm works, we define a *lexicographic ordering* of vectors. We say vector $(a_1a_2...a_n)$ *precedes* $(b_1b_2...b_k)$ *in lexicographic ordering* if $a_{k+1} = 1$, $b_{k+1} = 0$ for some k, $0 \le k \le n-1$, and $a_i = b_i$ for all $i \le k$. For example, (1 0 1 1 1 0) precedes (1 0 1 1 0 1) since they agree up to component 5, where the first vector has a 1, and the second a 0.

Example 28.

The lexicographic ordering of all 3-tuples over Z_2 is as follows.

1 1 1
1 1 0
1 0 1
1 0 0
0 1 1
0 1 0
0 0 1
0 0 0

We may view step-by-step decoding as a special version of standard array decoding in which a *lexicographic standard array* is constructed by specifically choosing as coset leader for each coset with non-unique minimum-weight vectors the vector of minimum weight in its coset which precedes all other such vectors with respect to lexicographical ordering. Table 1 of §3.6 is already in this form. Table 2 is not, since in the last coset of the given standard array, coset leader (001100) is preceded by the coset vector (100010). Changing the last row of Table 2 to be

100010 010110 111000 001011 001100 010001 111111 100101

satisfies the lexicographic constraint. Step-by-step decoding then corrects received vectors as standard array decoding does, with the correctable errors being precisely the coset leaders in the lexicographic standard array (exercise 59).

3.8 Weight Distribution and Weight Enumerators†

Let C be an $[n,M]$ code, and let A_i be the number of codewords of weight i in C. Then the vector $(A_o, A_1, \ldots, A_n)$ is called the *weight distribution* of C. In this section, we first see how knowledge of the weight distribution of a linear code can be used to determine the probability of improperly decoding received vectors. We then introduce the *weight enumerator* of a code, and discuss the well-known *MacWilliams identity*, which relates the weight distribution of a linear code to that of its dual code.

Let C be an (n,k)-code over $F = GF(q)$, and let the symbol error probability on the q-ary symmetric channel be p. Suppose the code C is used exclusively for error detection. The probability that errors go undetected by the decoder can then be determined as follows. Without loss of generality (since the code is linear), we may assume that the zero codeword is sent. The probability that a given codeword of weight i is received is then

† This section may be omitted without loss of continuity.

$$(\frac{p}{q-1})^i (1-p)^{n-i} , \; 0 \le i \le n .$$

Let the weight distribution for C be $(A_0, A_1, \ldots, A_n)$. The probability of errors going undetected is the probability that the error vector moves the transmitted codeword into another codeword, and this probability is therefore

$$\sum_{i=0}^{n} A_i (\frac{p}{q-1})^i (1-p)^{n-i} .$$

Hence knowledge of the weight distribution of the code is of use here.

Now consider the situation in which C is used to correct errors. Suppose C has distance $d = 2t + 1$, and the decoding scheme is that a received vector whose distance is at most t from a codeword is corrected to that codeword. (Note that this is an incomplete decoding scheme, since in general not all received vectors will be within distance t of a codeword, and hence not all received vectors will be decoded, correctly or otherwise, by the decoder.) The probability of correct decoding (see discussion in §1.3) is now exactly

$$\sum_{i=0}^{t} \binom{n}{i} p^i (1-p)^{n-i} ;$$

note that this probability is independent of the size of the alphabet. However, in this case, with both error correction and detection, the determination of the probability of the decoder making a decoding error is somewhat more difficult. We now derive this probability.

In order to proceed, we define a quantity $N(i,h;s)$. If there are not codewords of weight i, then $N(i,h;s) = 0$; otherwise, $N(i,h;s)$ is the number of vectors of weight h that are at distance s from a fixed codeword of weight i. $N(i,h;s)$ is independent of the given codeword of weight i (see exercise 98), and is hence well-defined. For $s \le t$, spheres of radius s about codewords are disjoint, and hence the number of vectors of weight h at distance exactly s from some codeword of weight i is

$$A_i \cdot N(i,h;s) .$$

Now received vectors which will be improperly decoded are those which lie within a sphere of radius t about some codeword other than that which was sent. Again, without loss of generality, assuming the zero codeword is sent, the probability of receiving a particular vector of weight h is

$$(\frac{p}{q-1})^h (1-p)^{n-h}.$$

Therefore the probability of receiving a vector of weight h which is distance exactly s ($s \le t$) from some codeword of weight i is

$$A_i \cdot N(i,h;s) \cdot (\frac{p}{q-1})^h (1-p)^{n-h} .$$

Now if the zero vector is sent, the received vector will be decoded to some codeword of fixed weight i with probability

$$\sum_{h=0}^{n} \sum_{s=0}^{t} A_i \cdot N(i,h;s) (\frac{p}{q-1})^h (1-p)^{n-h} .$$

If $i \geq 1$, a decoding error has occurred. Hence the probability of a decoding error is

$$\sum_{i=1}^{n} \sum_{h=0}^{n} \sum_{s=0}^{t} A_i \cdot N(i,h;s) (\frac{p}{q-1})^h (1-p)^{n-h} .$$

Again, knowledge of the weight distribution of the code is needed to apply this probability formula.

In addition to being of use in applying the above probability formulas, knowing the weight distribution $(A_0, A_1, \ldots, A_n)$ of a linear code allows one to determine the distance of the code immediately (recall Theorem 3.1). For any linear code with distance d, it is clear that $A_0 = 1$, $A_d \geq 1$ and $A_i = 0$ for $1 \leq i \leq d-1$, and hence if the A_i are known, d can be determined.

As one might expect, something as useful as the weight distribution is not easily found. For codes with only a small number of codewords, it is an easy matter to obtain the weight distribution by brute force - simply list all the codewords along with their weights, and tabulate the results; however, this is impractical for large codes, and in general, weight distributions are not known. To facilitate the study of these numbers, a special polynomial is introduced.

Definition. Let C be an (n,k)-code over F with weight distribution $(A_0, A_1, \ldots, A_n)$. The *weight enumerator* of C is defined to be

$$W_C(x,y) = \sum_{i=0}^{n} A_i x^{n-i} y^i .$$

For each vector $\mathbf{u} \in V_n(F)$, we also define $P(\mathbf{u}) = x^{n-w(\mathbf{u})} y^{w(\mathbf{u})}$ where x and y are indeterminates. Note that

$$\sum_{\mathbf{u}\in C} P(\mathbf{u}) = \sum_{i=0}^{n} A_i x^{n-i} y^i = W_C(x,y). \tag{3.3}$$

Example 29.

For $n = 2$,

$$P((0,0)) = x^2 \qquad P((0,1)) = xy$$
$$P((1,0)) = xy \qquad P((1,1)) = y^2 .$$

One of the most important results in this area is the MacWilliams identity which, as mentioned earlier, relates the weight enumerator of a linear code C to the weight enumerator of $C^{\perp}$. Hence knowing the weight enumerator of one of these codes enables one to determine the weight enumerator, and hence the weight distribution, of the other. This is useful in practice if, for example, one of C and $C^{\perp}$ is substantially smaller than the other, and the weight enumerator of the larger is required. One can determine the weight enumerator of the smaller code, perhaps even by exhaustive methods, and then, from this, obtain the (more difficult to determine directly) weight enumerator of the larger code by using the MacWilliams identity.

The MacWilliams identity is valid for any linear code over $F = GF(q)$, where q is a prime or prime number. We shall prove the case where q is a prime power. We actually give two proofs, one handling exclusively the case $q = 2$, and a second for the case $q \geq 2$, q prime. Although the proofs are very similar, and the second subsumes the first, we believe that inclusion of the first is warranted for those interested only in the binary case, as this first proof is somewhat simpler.

We begin with the binary case. Let $V_n = V_n(GF(2))$ denote the binary vector space of dimension n. For $\mathbf{u} \in V_n$, define

$$g_n(\mathbf{u}) = \sum_{\mathbf{v}\in V_n} (-1)^{\mathbf{u}\cdot\mathbf{v}} P(\mathbf{v}) ,$$

where $\mathbf{u}\cdot\mathbf{v}$ is the inner product as defined in §3.2. We use this definition in the following two lemmas, which are used to prove the MacWilliams identity.

Lemma 3.11. If C is a binary (n,k)-code, then

$$\sum_{\mathbf{u}\in C^{\perp}} P(\mathbf{u}) = \frac{1}{|C|} \sum_{\mathbf{u}\in C} g_n(\mathbf{u}) .$$

Proof.

$$\sum_{\mathbf{u}\in C} g_n(\mathbf{u}) = \sum_{\mathbf{u}\in C} \sum_{\mathbf{v}\in V_n} (-1)^{\mathbf{u}\cdot\mathbf{v}} P(\mathbf{v}) = \sum_{\mathbf{v}\in V_n} P(\mathbf{v})\, S(\mathbf{v})$$

where

$$S(\mathbf{v}) = \sum_{\mathbf{u}\in C} (-1)^{\mathbf{u}\cdot\mathbf{v}} , \ \mathbf{v}\in V_n .$$

We first prove that

$$S(\mathbf{v}) = \begin{cases} |C| & \text{if } \mathbf{v}\in C^{\perp} \\ 0 & \text{if } \mathbf{v}\notin C^{\perp} . \end{cases}$$

This is most easily done by using some group arguments. For $\mathbf{v}\in V_n$, let

$$C_0(\mathbf{v}) = \{\mathbf{u}\in C :\ \mathbf{u}\cdot\mathbf{v} = 0\} .$$

It is easily checked (and the reader is encouraged to do so) that $C_0(\mathbf{v})$ is a subgroup of C (and in this case, of course a subspace also). Now if $\mathbf{w}_1, \mathbf{w}_2 \in C$ and $\mathbf{w}_1\cdot\mathbf{v} = 1$ and $\mathbf{w}_2\cdot\mathbf{v} = 1$ then $(\mathbf{w}_1 - \mathbf{w}_2)\cdot\mathbf{v} = 0$ and $\mathbf{w}_1 - \mathbf{w}_2 \in C_0(\mathbf{v})$. This implies that

$$C_1(\mathbf{v}) = \{\mathbf{u}\in C :\ \mathbf{u}\cdot\mathbf{v} = 1\}$$

is either the empty set or a coset of $C_0(\mathbf{v})$, with $C_1(\mathbf{v}) = \varnothing$ if and only if $\mathbf{v}\in C^{\perp}$. Now for $\mathbf{v}\in C^{\perp}$,

$$S(\mathbf{v}) = \sum_{\mathbf{u}\in C_0(\mathbf{v})} (-1)^0 = |C| , \qquad \mathbf{v}\in C^{\perp} .$$

For $\mathbf{v}\notin C^{\perp}$, then $|C_0(\mathbf{v})| = |C(\mathbf{v})|$, and since $C = C_0(\mathbf{v}) \cup C_1(\mathbf{v})$, we have

$$S(\mathbf{v}) = \sum_{\mathbf{u}\in C_0(\mathbf{v})} (-1)^{\mathbf{u}\cdot\mathbf{v}} + \sum_{\mathbf{u}\in C_1(\mathbf{v})} (-1)^{\mathbf{u}\cdot\mathbf{v}} = 0 , \ \ \mathbf{v}\notin C^{\perp} .$$

Returning to $\sum_{\mathbf{u}\in C} g_n(\mathbf{u})$, we have

$$\sum_{\mathbf{u}\in C} g_n(\mathbf{u}) = \sum_{\mathbf{v}\in C^\perp} P(\mathbf{v})S(\mathbf{v}) + \sum_{\mathbf{v}\notin C^\perp} P(\mathbf{v})S(\mathbf{v})$$

$$= |C| \sum_{\mathbf{v}\in C^\perp} P(\mathbf{v}) .$$

The result now follows. □

Lemma 3.12. Let $\mathbf{u} \in V_n = V_n(GF(2))$. Then

$$g_n(\mathbf{u}) = (x+y)^{n-w(\mathbf{u})}(x-y)^{w(\mathbf{u})} .$$

Proof.

The proof will be by induction on n. By definition,

$$g_n(\mathbf{u}) = \sum_{\mathbf{v}\in V_n} (-1)^{\mathbf{u}\cdot\mathbf{v}} P(\mathbf{v}) , \quad \mathbf{u} \in V_n .$$

For $n = 1$,

$$g_n(\mathbf{u}) = (-1)^{\mathbf{u}\cdot 0} P((0)) + (-1)^{\mathbf{u}\cdot 1} P((1))$$

$$= \begin{cases} x+y , & \mathbf{u} = (0) \\ x-y , & \mathbf{u} = (1) \end{cases}$$

$$= \begin{cases} (x+y)^{1-w((0))} (x-y)^{w((0))} , & \mathbf{u} = 0 \\ (x+y)^{1-w((1))} (x-y)^{w((1))} , & \mathbf{u} = 1 \end{cases}$$

$$= (x+y)^{1-w(\mathbf{u})} (x-y)^{w(\mathbf{u})} , \quad \mathbf{u} \in V_1 .$$

Hence the statement is true for $n = 1$. Now as inductive hypothesis, assume that the statement is true for $n = k$, and then consider the case $n = k + 1$. Let

$$\mathbf{u} = (u_1, u_2, \ldots, u_{k+1}), \quad \mathbf{v} = (v_1, v_2, \ldots, v_{k+1}) ,$$

$$\mathbf{u}' = (u_1, u_2, \ldots, u_k) , \quad \mathbf{v}' = (v_1, v_2, \ldots, v_k) .$$

Then

$$g_{k+1}(\mathbf{u}) = \sum_{\substack{\mathbf{v}\in V_{k+1}\\ v_{k+1}=0}} (-1)^{\mathbf{u}\cdot\mathbf{v}} P(\mathbf{v}) + \sum_{\substack{\mathbf{v}\in V_{k+1}\\ v_{k+1}=1}} (-1)^{\mathbf{u}\cdot\mathbf{v}} P(\mathbf{v})$$

$$= \sum_{\mathbf{v}'\in V_k} (-1)^{\mathbf{u}'\cdot\mathbf{v}'} x^{k+1-w(\mathbf{v}')} y^{w(\mathbf{v}')} + \sum_{\mathbf{v}'\in V_k} (-1)^{\mathbf{u}'\cdot\mathbf{v}'+u_{k+1}} x^{k-w(\mathbf{v}')} y^{w(\mathbf{v}')+1}$$

$$= x g_k(\mathbf{u}') + y(-1)^{u_{k+1}} g_k(\mathbf{u}')$$

$$= g_k(\mathbf{u}')(x + y(-1)^{u_{k+1}})$$

which now by the inductive hypothesis is

$$= (x+y)^{k-w(\mathbf{u}')}(x-y)^{w(\mathbf{u}')}(x+y(-1)^{u_{k+1}})$$

which is easily seen (consider $u_{k+1} = 0$, $u_{k+1} = 1$) to be

$$= (x+y)^{k+1-w(\mathbf{u})}(x-y)^{w(\mathbf{u})} .$$

Hence the statement is true for $n = k + 1$, to complete the induction. □

We now use these two lemmas to prove the important result by MacWilliams (for binary linear codes) that shows how, given the weight enumerator for a code, one can obtain the weight enumerator of the dual code.

Theorem 3.13. (MacWilliams identity over $GF(2)$). If C is a binary (n,k)-code with dual code $C^\perp$, then

$$W_{C^\perp}(x,y) = \frac{1}{2^k} W_C(x+y, x-y) .$$

Proof.

Let the weight distribution of C be $(A_0, A_1, \ldots, A_n)$. Then

$$\sum_{\mathbf{u}\in C^\perp} P(\mathbf{u}) = \frac{1}{|C|} \sum_{\mathbf{u}\in C} g_n(\mathbf{u}) \qquad \text{by Lemma 3.11}$$

$$= \frac{1}{|C|} \sum_{\mathbf{u}\in C} (x+y)^{n-w(\mathbf{u})}(x-y)^{w(\mathbf{u})} \qquad \text{by Lemma 3.12}$$

$$= \frac{1}{|C|} \sum_{i=0}^{n} A_i (x+y)^{n-i}(x-y)^i$$

$$= \frac{1}{|C|} W_C\,(x+y, x-y) \qquad \text{by equation (3.3).}$$

Hence, by (3.3) again, we have

$$W_{C^\perp}(x+y) = \frac{1}{|C|} W_C(x+y, x-y),$$

and $|C| = 2^k$. □

Example 30.

Consider the code given in example 10. C is generated by G and has eight codewords. $C^\perp$ is generated by H and, in this case, also has eight codewords. We list the codewords of C and $C^\perp$ along with their weights.

C	wt	$C^\perp$	wt
(0 0 0 0 0 0)	0	(0 0 0 0 0 0)	0
(1 0 0 1 1 0)	3	(1 0 1 1 0 0)	3
(0 1 0 0 1 1)	3	(1 1 0 0 1 0)	3
(0 0 1 1 0 1)	3	(0 1 1 0 0 1)	3
(1 1 0 1 0 1)	4	(0 1 1 1 1 0)	4
(0 1 1 1 1 0)	4	(1 0 1 0 1 1)	4
(1 0 1 0 1 1)	4	(1 1 0 1 0 1)	4
(1 1 1 0 0 0)	3	(0 0 0 1 1 1)	3

The weight distribution of C is (1,0,0,4,3,0,0), and hence the weight enumerator of C is

$$W_C(x,y) = x^6 + 4x^3y^3 + 3x^2y^4.$$

From the vectors of $C^\perp$ listed above, the weight enumerator for $C^\perp$ can also be written down. This latter weight enumerator can also be determined as follows. Compute

$$\begin{aligned}
W_C(x+y, x-y) &= (x+y)^6 + 4(x+y)^3(x-y)^3 + 3(x+y)^2(x-y)^4 \\
&= (x+y)^3((x+y)^3 + 4(x-y)^3) + 3(x+y)^2(x-y)^4 \\
&= (x+y)^3[x^3 + 3x^2y + 3xy^2 + y^3 + 4x^3 - 12x^2y \\
&\qquad + 12xy^2 - 4y^3] + 3(x+y)^2(x-y)^4 \\
&= (x+y)^3[5x^3 - 9x^2y + 15xy^2 - 3y^3] + 3(x+y)^2(x-y)^4 \\
&= (x+y)^2[(x+y)(5x^3 - 9x^2y + 15xy^2 - 3y^3) + 3(x-y)^4]
\end{aligned}$$

$$= (x+y)^2[8x^4 - 16x^3y + 24x^2y^2]$$
$$= 8x^6 + 32x^3y^3 + 24x^2y^4.$$

Now by Theorem 3.13,

$$W_{C^\perp}(x,y) = \frac{1}{8}W_C(x+y,x-y) = x^6 + 4x^3y^3 + 3x^2y^4 .$$

In this case, C and $C^\perp$ have the same weight enumerator. Note that $C \neq C^\perp$.

We now consider the slightly more general case $q = p \geq 2$, p a prime number. Let α be a *primitive complex p^{th} root of unity*, ie. $\alpha = e^{2\pi i/p}$ where $e^{2\pi i/p}$ denotes $\cos(2\pi/p) + i\sin(2\pi/p)$, and $i^2 = -1$. (Note that for $p = 2$, $\alpha = e^{\pi i} = -1$, yielding the proof in the binary case). Let $V_n(F)$ be the n-dimensional vector space over $GF(p)$. As before, define

$$P(\mathbf{u}) = x^{n-w(\mathbf{u})}y^{w(\mathbf{u})}$$

with x and y indeterminates, and now define

$$g_n(\mathbf{u}) = \sum_{\mathbf{v}\in V_n(F)} \alpha^{\mathbf{u}\cdot\mathbf{v}}P(\mathbf{v}), \quad \mathbf{u} \in V_n(F) \tag{3.4}$$

Lemma 3.14. If C is an (n,k)-code over $F = GF(p)$, then

$$\sum_{\mathbf{u}\in C^\perp} P(\mathbf{u}) = \frac{1}{|C|}\sum_{\mathbf{u}\in C} g_n(\mathbf{u}) .$$

Proof.

$$\sum_{\mathbf{u}\in C} g_n(\mathbf{u}) = \sum_{\mathbf{u}\in C}\sum_{\mathbf{v}\in V_n(F)} \alpha^{\mathbf{u}\cdot\mathbf{v}}P(\mathbf{v}) = \sum_{\mathbf{v}\in V_n(F)} P(\mathbf{v})S(\mathbf{v})$$

where now

$$S(\mathbf{v}) = \sum_{\mathbf{u}\in C} \alpha^{\mathbf{u}\cdot\mathbf{v}} , \quad \mathbf{v} \in V_n(F) .$$

As in the proof of Lemma 3.11, we first prove that

$$S(\mathbf{v}) = \begin{cases} |C| & \text{if } \mathbf{v} \in C^\perp \\ 0 & \text{if } \mathbf{v} \notin C^\perp \end{cases} .$$

Let

$$C_0(\mathbf{v}) = \{\mathbf{u} \in C : \mathbf{u}\cdot\mathbf{v} = 0\} , \quad \mathbf{v} \in V_n(F) .$$

Then $C_0(\mathbf{v})$ is again a subgroup of C, and is in fact a subspace. Now as before, two elements $\mathbf{w}_1, \mathbf{w}_2 \in C$ are in the same coset of $C_0(\mathbf{v})$ in C if and only if $\mathbf{w}_1 - \mathbf{w}_2 \in C_0(\mathbf{v})$, that is, if and only if $(\mathbf{w}_1 - \mathbf{w}_2)\cdot\mathbf{v} = 0$ or $\mathbf{w}_1\cdot\mathbf{v} = \mathbf{w}_2\cdot\mathbf{v}$. This implies that

$$C_i(\mathbf{v}) = \{\mathbf{u} \in C : \mathbf{u}\cdot\mathbf{v} = i\} , \quad 0 \le i \le p-1$$

is a coset of $C_0(\mathbf{v})$ if and only if $C_0(\mathbf{v}) \ne C$, which is true if and only if $\mathbf{v} \notin C^\perp$. If $\mathbf{v} \notin C^\perp$ then $|C_0(\mathbf{v})| = |C_i(\mathbf{v})|$ for $0 \le i \le p-1$, and $C = C_0(\mathbf{v}) \cup C_1(\mathbf{v}) \cup \cdots \cup C_{p-1}(\mathbf{v})$ is a disjoint union of cosets, so that

$$\begin{aligned} S(\mathbf{v}) &= \sum_{i=0}^{p-1} \sum_{\mathbf{u}\in C_i(\mathbf{v})} \alpha^{\mathbf{u}\cdot\mathbf{v}} \\ &= \sum_{i=0}^{p-1} \sum_{\mathbf{u}\in C_i(\mathbf{v})} \alpha^{i} \\ &= \sum_{i=0}^{p-1} p^{k-1}\alpha^{i} \\ &= p^{k-1}\left(\frac{1-\alpha^p}{1-\alpha}\right) \end{aligned}$$

using the formula for a geometric progression; note $\alpha \ne 1$. But now $\alpha^p = 1$ implies $S(\mathbf{v}) = 0$ for $\mathbf{v} \notin C^\perp$. Furthermore, for $\mathbf{v} \in C^\perp$,

$$S(\mathbf{v}) = \sum_{\mathbf{u}\in C} \alpha^{\mathbf{u}\cdot\mathbf{v}} = \sum_{\mathbf{u}\in C} \alpha^0 = |C| .$$

Now

$$\sum_{\mathbf{u}\in C} g_n(\mathbf{u}) = \sum_{\mathbf{v}\in V_n(F)} P(\mathbf{v})\, S(\mathbf{v})$$

$$= \sum_{\mathbf{v}\in C^{\perp}} P(\mathbf{v})\,S(\mathbf{v}) + \sum_{\mathbf{v}\notin C^{\perp}} P(\mathbf{v})\,S(\mathbf{v})$$

$$= |C| \sum_{\mathbf{v}\in C^{\perp}} P(\mathbf{v}) .$$

The result follows. □

Lemma 3.15. Let $\mathbf{u} \in V_n(F)$ where $F = GF(p)$ and p is prime. Then

$$g_n(\mathbf{u}) = (x + (p-1)y)^{n-w(\mathbf{u})}(x-y)^{w(\mathbf{u})}$$

Proof.

The proof is by induction on n. By definition from (3.4),

$$g_n(\mathbf{u}) = \sum_{\mathbf{v}\in V_n} \alpha^{\mathbf{u}\cdot\mathbf{v}} P(\mathbf{v}), \quad \mathbf{u} \in V_n(F) .$$

For $n = 1$,

$$g_i(\mathbf{u}) = \begin{cases} \sum_{i=0}^{p-1} P((i)), & \mathbf{u} = 0 \\ \sum_{i=0}^{p-1} \alpha^i P((i)), & \mathbf{u} \neq 0 \end{cases}$$

$$= \begin{cases} x + (p-1)y, & \mathbf{u} = 0 \\ x - y, & \mathbf{u} \neq 0 \end{cases}$$

where the last line follows since

$$1 + \alpha + \cdots + \alpha^{p-1} = \frac{1-\alpha^p}{1-\alpha} = 0 \quad \text{implies} \quad \sum_{i=1}^{p-1} \alpha^i = -1 .$$

In both cases,

$$g_1(\mathbf{u}) = (x + (p-1)y)^{n-w(\mathbf{u})}(x-y)^{w(\mathbf{u})},$$

and the statement holds for $n = 1$. Now assume the statement is true for $n = k$, and consider $n = k + 1$. Let

$$\mathbf{u} = (u_1, u_2, \ldots, u_{k+1}), \qquad \mathbf{v} = (v_1, v_2, \ldots, v_{k+1}),$$

$$\mathbf{u}' = (u_1, u_2, \ldots, u_k), \qquad \mathbf{v}' = (v_1, v_2, \ldots, v_k).$$

Then

$$g_{k+1}(\mathbf{u}) = \sum_{\substack{\mathbf{v} \in V_{k+1}(F) \\ v_{k+1} = 0}} \alpha^{\mathbf{u} \cdot \mathbf{v}} P(\mathbf{v}) + \sum_{\substack{\mathbf{v} \in V_{k+1}(F) \\ v_{k+1} \neq 0}} \alpha^{\mathbf{u} \cdot \mathbf{v}} P(\mathbf{v})$$

$$= \sum_{\mathbf{v}' \in V_k(F)} \alpha^{\mathbf{u}' \cdot \mathbf{v}'} x^{k+1-w(\mathbf{v}')} y^{w(\mathbf{v}')} + \sum_{i=1}^{p-1} \sum_{\mathbf{v}' \in V_k(F)} \alpha^{\mathbf{u}' \cdot \mathbf{v}' + i u_{k+1}} x^{k-w(\mathbf{v}')} y^{w(\mathbf{v}')+1}$$

$$= x\, g_k(\mathbf{u}') + y \sum_{i=1}^{p-1} \alpha^{i u_{k+1}} g_k(\mathbf{u}')$$

$$= g_k(\mathbf{u}')\, (x + y \sum_{i=1}^{p-1} \alpha^{i u_{k+1}})$$

which now, applying the inductive hypothesis, becomes

$$= (x + (p-1)y)^{k-w(\mathbf{u}')}(x-y)^{w(\mathbf{u}')}\, (x + y \sum_{i=1}^{p-1} \alpha^{i u_{k+1}})$$

and is easily verfied (consider $u_{k+1} = 0$, $u_{k+1} \neq 0$) to be

$$= (x + (p-1)y)^{k+1-w(\mathbf{u})}(x - y)^{w(\mathbf{u})}.$$

This proves the statement true for $n = k + 1$, completing the induction. □

We can now state the MacWilliams identity of Theorem 3.13 for codes over $F = GF(p)$.

Theorem 3.16 (MacWilliams identity over $GF(p)$). If C is an (n,k)-code over $F = GF(p)$, p a prime (so that $|C| = p^k$), with dual code $C^{\perp}$, then

$$W_{C^{\perp}}(x,y) = \frac{1}{|C|}\, W_C(x + (p-1)y\,,\; x-y).$$

Proof.

The proof follows from Lemmas 3.14 and 3.15, just as the proof of Theorem 3.13 followed from Lemmas 3.11 and 3.12. □

Again we note that Theorem 3.16 and its proof eliminate the need for Theorem 3.13 (and Lemmas 3.11 and 3.12), although for reasons previously mentioned, we feel strongly that their inclusion is worthwhile.

We now apply Theorem 3.16 to some of our previous examples of codes over prime fields other than $GF(2)$.

Example 31.

Example 11 constructs a (5,2)-code C over $GF(3)$ having generator matrix

$$G = \begin{bmatrix} 2\,0\,2\,1\,0 \\ 1\,1\,0\,0\,1 \end{bmatrix}.$$

The code C has 9 vectors as listed in example 11. From this list, the weight distribution of C is seen to be (1,0,0,4,2,2), and hence the weight enumerator for C is given by

$$W_C(x,y) = x^5 + 4x^2y^3 + 2xy^4 + 2y^5 .$$

From Theorem 3.16 we have

$$W_{C^\perp}(x,y) = \frac{1}{9}\, W_C\,(x{+}2y,\, x{-}y)$$

$$= \frac{1}{9}(9x^5 + 36x^3y^2 + 72x^2y^3 + 108xy^4 + 18y^5)$$

$$= x^5 + 4x^3y^2 + 8x^2y^3 + 12xy^4 + 2y^5 .$$

Note that $C^\perp$ is a (5,3)-code with $3^3 = 27$ codewords, and hence determining the weight enumerator for $C^\perp$ by the naive method of listing all codewords and determining their weights becomes more difficult than doing so for C. As a quick check, note that the sum of the coefficients in $W_{C^\perp}(x,y)$ here is 27.

Example 32.

Example 13 constructed a (6,2)-code C over $GF(7)$ having distance 5. The dual code $C^\perp$ is a (6,4)-code having 7^4 codewords. It would be tedious to determine the weight distribution and construct the enumerator for $C^\perp$ directly. Hence we use Theorem 3.16 to find this weight enumerator indirectly. We first require the weight enumerator for C. A generator matrix for C can be derived from the matrix H of example 13 as

$$G = \begin{bmatrix} 1\,0\,4\,2\,3\,6 \\ 0\,1\,4\,6\,5\,2 \end{bmatrix}.$$

Consider the following codewords in C.

$$\begin{matrix} (1\,1\,1\,1\,1\,1) \\ (1\,2\,5\,0\,6\,3) \\ (1\,3\,2\,6\,4\,5) \\ (1\,4\,6\,5\,2\,0) \\ (1\,5\,3\,4\,0\,2) \\ (1\,6\,0\,3\,5\,4) \end{matrix}.$$

Taking these six codewords, and the two codewords which are rows of G yields eight codewords; the remaining non-zero codewords of C are obtained by taking all non-zero multiples (over $GF(7)$) of these eight. Since a non-zero multiple of a codeword has the same weight as that codeword, from these eight codewords we can obtain the weights of all 49 codewords in C. The weight enumerator of C is hence found to be

$$W_C(x,y) = x^6 + 36xy^5 + 12y^6 .$$

Applying Theorem 3.16 now gives

$$\begin{aligned} W_{C^\perp}(x,y) &= \frac{1}{|C|}\, W_C(x+6y, x-y) \\ &= \frac{1}{49}\,(49x^6 + 5880x^3y^3 + 17640x^2y^4 + 47628xy^5 \\ &= x^6 + 120x^3y^3 + 360x^2y^4 + 972xy^5 + 948y^6 . \end{aligned}$$

The statement of the MacWilliams identity for the general case of any field $F = GF(q)$, q a prime or a power of a prime number, is identical to that of Theorem 3.16 with p and $GF(p)$ from Theorem 3.16 replaced by q and $GF(q)$. The proof of this result requires a more general definition for $g_n(\mathbf{u})$, and we do not pursue the matter here.

3.9 Exercises

1. (a) Find a parity-check matrix for the code corresponding to each of the following generator matrices. G_1, G_3 and G_4 are over $GF(2)$; G_2 is over $GF(3)$.

$$G_1 = \begin{bmatrix} 1&0&0&1&1 \\ 1&1&1&0&1 \\ 0&0&1&1&1 \end{bmatrix} \quad G_2 = \begin{bmatrix} 1&2&1&1&2&1 \\ 0&1&0&1&1&1 \\ 1&0&2&1&1&1 \\ 0&1&0&2&1&0 \end{bmatrix}$$

$$G_3 = \begin{bmatrix} 1&0&1&0&1&1 \\ 1&1&0&1&0&1 \\ 1&1&1&1&0&1 \\ 1&0&0&0&1&1 \end{bmatrix} \quad G_4 = \begin{bmatrix} 1&1&1&0&0&1&0&0&0&0&0 \\ 0&0&1&1&1&0&1&0&0&0&0 \\ 0&1&1&0&1&0&0&1&0&0&0 \\ 0&1&0&1&1&0&0&0&1&0&0 \\ 0&1&1&1&0&0&0&0&0&1&0 \\ 1&1&1&1&1&0&0&0&0&0&1 \end{bmatrix}$$

   (b) Is the code with generator matrix G_3 very useful? Explain.

   (c) Prove that in the code with generator matrix G_4, all codewords have even weight.

2. (a) Determine the distance of the codes specified by the following parity-check matrices. H_1 and H_4 are over $GF(2)$; H_2 is over $GF(7)$, and H_3 is over $GF(31)$.

$$H_1 = \begin{bmatrix} 1&0&0&1&0&0&0&1&1&0&1&1&0&1 \\ 0&0&1&0&1&0&1&1&1&0&0&1&1&1 \\ 0&0&1&0&0&1&0&0&0&1&1&1&0&1 \\ 0&1&0&1&0&0&1&1&0&1&0&1&1&0 \end{bmatrix} \quad H_2 = \begin{bmatrix} 2&1&0&6&1&0&0 \\ 0&1&3&0&4&3&2 \\ 4&0&4&6&0&3&5 \end{bmatrix}$$

$$H_3 = \begin{bmatrix} 1&1&0&0&0&1&0 \\ 0&1&1&1&0&0&0 \\ 0&1&0&1&0&1&1 \end{bmatrix} \quad H_4 = \begin{bmatrix} 1&0&0&0&1&0&1&1 \\ 0&1&0&0&0&1&1&1 \\ 0&0&1&0&1&1&0&1 \\ 0&0&0&1&1&1&1&0 \end{bmatrix}$$

   (b) Of the codes corresponding to these parity-check matrices, which are useful for error correction? Which are useful for error detection?

3. Let C be the code generated by the matrix

$$G = \begin{bmatrix} 1 & 1 & 1 & 0 & 1 & 1 \\ 0 & 1 & 0 & 0 & 1 & 1 \\ 1 & 0 & 1 & 1 & 0 & 1 \\ 0 & 1 & 1 & 1 & 0 & 1 \end{bmatrix}$$

(a) Determine the number of codewords in C.

(b) Find a parity-check matrix for C.

(c) Determine the distance d of C.

(d) Determine the number of errors C can (i) correct, and (ii) detect.

4. Let H be a matrix such that the span of its rows generates the orthogonal complement for a binary (n,k)-code C, where

$$H = \begin{bmatrix} 1 & 1 & 0 & 1 & 0 & 0 & 1 \\ 0 & 1 & 1 & 1 & 1 & 0 & 0 \\ 1 & 0 & 0 & 1 & 1 & 1 & 0 \\ 0 & 0 & 1 & 1 & 0 & 1 & 1 \end{bmatrix}$$

(a) Determine n, k, the number of codewords M and the distance d of C.

(b) Find a generator matrix G for C.

(c) Prove that $C^{\perp} \subset C$.

(d) Decode the following received vectors; state any assumptions made.
(i) $\mathbf{r}_1 = (1\,1\,1\,0\,1\,0\,1)$ (ii) $\mathbf{r}_2 = (0\,1\,1\,0\,1\,0\,1)$ (iii) $\mathbf{r}_3 = (1\,1\,1\,0\,1\,1\,1)$
(iv) $\mathbf{r}_4 = (1\,1\,1\,0\,0\,1\,1)$ (v) $\mathbf{r}_5 = (1\,1\,1\,0\,0\,0\,1)$

5. Determine the smallest integer n for which a single-error correcting binary linear code of block length n carrying 62 bits of information per codeword can be constructed.

6. A source encoder assigns binary 4-tuples to symbols as follows:

space	→	0011	M	→	0010
E	→	1100	P	→	1101
H	→	0110	U	→	1010
L	→	0100	A	→	0101

Given a message 4-tuple $\mathbf{m}$, a channel encoder uses the matrix G to construct the codeword $\mathbf{m}G$, where

$$G = \begin{bmatrix} 0&0&0&1&1&1&0&1 \\ 1&0&0&1&0&1&1&0 \\ 0&1&0&1&1&0&1&0 \\ 0&0&1&0&1&1&1&0 \end{bmatrix}.$$

Decode the following received sequence of bits, correcting each vector to the nearest codeword.

11010110 10001011 10011000 10101101 11110100 01000110 10100101

7. Assign characters to 4-tuples as follows.

space	→	0000	M	→	1001
A	→	0001	N	→	1010
B	→	0010	O	→	0101
C	→	0100	R	→	0111
D	→	1000	S	→	1110
E	→	0011	T	→	1011
F	→	0110	U	→	1101
G	→	1100	Y	→	1111

These information 4-tuples are encoded using the binary generator matrix

$$G = \begin{bmatrix} 1&1&1&0&0&0&0 \\ 1&0&0&1&1&0&0 \\ 0&1&0&1&0&1&0 \\ 1&1&0&1&0&0&1 \end{bmatrix}.$$

Assuming that at most one error will occur in any 7 bit transmission, decode the following received sequences.

(a) 1110001 1100011 1001100 0100111 0110000 1000011 1000000 0101011

(b) 1110011 1000111 1111001 0011111 0010110 1000000 1100110 0100111 0001111 0000001 1100111 1100011 1101001 1001111 0010110

(c) 1010110 1000011 0000011 0001000 1101111 0000101 1110101 0000000 0011110 0100101 1100101 1011010

8. Prove that given an (n,k)-code C, there exists a generator matrix G for C or for an equivalent code C' such that $G = [I_k\, A]$.

9. (a) Prove that the even weight vectors in a binary linear code form an additive group.

   (b) Prove that in a binary linear code either all codewords have even weight or exactly half have even weight and half have odd weight. (Hint: consider cosets of the group from (a))

10. Prove that in a binary linear code, either all codewords begin with 0, or exactly half begin with 0 and half begin with 1.

11. Prove that if the rows of a generator matrix G for a binary linear code all have even weight, then all codewords have even weight.

12. Let the matrix whose rows span the orthogonal complement for a binary (5,2)-code C be

$$H = \begin{bmatrix} 1 & 0 & 0 & 1 & 0 \\ 0 & 1 & 0 & 1 & 1 \\ 0 & 0 & 1 & 0 & 1 \end{bmatrix}$$

   (a) Determine whether or not C is a perfect code.

   (b) Construct a lexicographic standard array for C.

   (c) Construct a one-to-one correspondence between coset leaders in (b) and syndromes.

   (d) Decode the following received vectors.
   (i) $\mathbf{r}_1 = (1\ 0\ 1\ 0\ 1)$ (ii) $\mathbf{r}_2 = (0\ 1\ 1\ 1\ 1)$
   (iii) $\mathbf{r}_3 = (1\ 1\ 1\ 1\ 1)$ (iv) $\mathbf{r}_4 = (1\ 1\ 1\ 0\ 0)$
   (v) $\mathbf{r}_5 = (1\ 0\ 1\ 1\ 1)$

13. Let a parity-check matrix for a (4,2)-code C over Z_3 be $H = \begin{bmatrix} 1 & 0 & 2 & 1 \\ 0 & 1 & 1 & 1 \end{bmatrix}$.

   (a) Determine whether or not C is a perfect code.

   (b) Construct a lexicographic standard array for C.

   (c) Construct a one-to-one correspondence between coset leaders in (b) and syndromes.

   (d) Decode the following received vectors.
   (i) $\mathbf{r}_1 = (2\ 2\ 1\ 1)$ (ii) $\mathbf{r}_2 = (2\ 2\ 2\ 1)$ (iii) $\mathbf{r}_3 = (2\ 2\ 2\ 2)$

14. Consider the binary code generated by the matrix

$$G = \begin{bmatrix} 1 & 0 & 0 & 1 & 1 & 0 \\ 0 & 1 & 0 & 0 & 1 & 1 \\ 0 & 0 & 1 & 1 & 0 & 1 \end{bmatrix}.$$

(a) **Construct a 1-1 correspondence between syndromes and weights of the lightest vectors in the corresponding cosets.**

(b) **Decode the following received vectors, using step-by-step decoding:**
(i) $\mathbf{r}_1 = (100010)$ **(ii)** $\mathbf{r}_2 = (100001)$ **(iii)** $\mathbf{r}_3 = (010100)$ **(iv)** $\mathbf{r}_4 = (111100)$

(c) **Which of the vectors in (b) were decoded to a *unique* closest codeword?**

15. **Consider the binary (5,2)-code C with parity-check matrix** $H = \begin{bmatrix} 1 & 1 & 1 & 0 & 0 \\ 0 & 1 & 0 & 1 & 0 \\ 1 & 0 & 0 & 0 & 1 \end{bmatrix}$

(a) **Display a generator matrix G for C.**

(b) **Construct a (lexicographic) standard array for C.**

(c) **Construct the one-to-one correspondence between coset leaders and syndromes.**

(d) **Using this correspondence and syndrome decoding, decode**
(i) $\mathbf{r}_1 = (11101)$ **(ii)** $\mathbf{r}_2 = (11011)$

(e) **Construct a table of syndromes and the weights of the corresponding coset leaders.**

(f) **Decode $\mathbf{r}_3 = (11100)$ using step-by-step decoding.**

16. **A binary (n,k)-code C has generator matrix**

$$G = \begin{bmatrix} 1 & 0 & 1 & 1 & 1 & 0 & 0 & 0 \\ 1 & 1 & 0 & 0 & 1 & 1 & 0 & 0 \\ 0 & 1 & 0 & 1 & 1 & 0 & 1 & 0 \\ 0 & 0 & 1 & 1 & 0 & 0 & 1 & 1 \end{bmatrix}.$$

(a) **Determine n, k, and the number of codewords M in C.**

(b) **Construct a parity-check matrix H, and determine the distance d of C.**

(c) **Determine the error pattern associated with the vector $\mathbf{r} = (1111\ 0010)$, assuming at most one error has occurred.**

(d) **In a standard array for C, determine the number of cosets with coset leaders of weight greater than or equal to 2.**

17. Let H be a parity check matrix for a binary (n,k)-code C, with $n \geq 4$. Suppose the columns of H are distinct and all columns have odd weight. Prove that C has distance $d \geq 4$.

18. Let C be an (n,k)-code over $GF(q)$, and assume that no component is identically 0 in all codewords.

    (a) Show that if all the code vectors in C are arranged as the rows of a matrix, then each field element appears precisely q^{k-1} times in each column. (Hint: show that the set of all codewords with 0's in a particular component forms a subspace, and then consider cosets of this subspace)

    (b) Using (a), show that the sum of weights of all codewords in C is $n(q-1)q^{k-1}$.

19. (a) Let C be a binary (n,k)-code with generator matrix G whose columns are precisely the 2^k-1 non-zero binary k-tuples. Prove that all non-zero codewords have weight 2^{k-1}.

    (b) More generally, let C be an (n,k)-code over $GF(q)$ with generator matrix G whose columns are precisely the q^k-1 non-zero k-tuples over $GF(q)$. Prove that all non-zero codewords have weight q^k-q^{k-1}.

20. Show that if the columns of the generator matrix G of a code consist of $(q^k-1)/(q-1)$ vectors, none of which is a scalar multiple of another, then all non-zero codewords have weight q^{k-1}. (Hint: see problem above)

21. Let H be the parity-check matrix for a linear code. Show that the coset whose syndrome is v contains a vector of weight w if and only if some linear combinations of w columns of H equals v.

22. Let C be a binary $(n,n-m)$-code, $2^{m-1} \leq n < 2^m$, whose $m \times n$ parity-check matrix H has as column i the binary representation of i, $1 \leq i \leq n$. Prove there exists a vector of weight 2 or less in every coset of C. (Hint: recall that each coset has a unique syndrome)

23. Consider the parity-check matrix

$$H = \begin{bmatrix} 1&0&0&0&1&0&1&1&1&1 \\ 1&1&0&0&0&0&1&0&0&1 \\ 0&1&1&0&1&1&1&0&0&1 \\ 1&1&0&1&0&1&1&0&1&0 \end{bmatrix}.$$

    (a) Determine the values n,k and d such that H is the parity-check matrix for a binary (n,k,d)-code C.

(b) Find a generator matrix G for C.

(c) Set up a one-to-one correspondence between syndromes and coset leaders. (Use the parity-check matrix H given above.)

(d) Decode each of the following vectors.
(i) $\mathbf{r}_1 = (11101\ 10000)$ (ii) $\mathbf{r}_2 = (01001\ 10010)$ (iii) $\mathbf{r}_3 = (10111\ 10010)$.

(e) Determine whether or not this code is perfect.

24. Let C be the binary code in the problem above. Extend each codeword by appending an 11th bit. If $\mathbf{c} \in C$ has odd weight, let the new bit be 1, and 0 otherwise. This is called adding an *overall parity-check bit.*

(a) Determine the distance in the new code.

(b) Construct a parity-check matrix H' for the new code by appending an appropriate row and column to the parity-check matrix H.

25. Let d be the weight of a linear code C, and let $t = \lfloor (d-1)/2 \rfloor$. Suppose d is even. Show that there are two vectors of weight $t+1$ in some coset of C.

26. Find the largest n such that there is a binary code with distance $d = 5$ and at most 8 check symbols.

27. Define the probability of error, P_{error}, for a particular decoding scheme to be the probability that the decoder decodes a received vector erroneously. Suppose we are decoding using a standard array. Show that

$$P_{error} = 1 - \sum_{i=0}^{n} \alpha_i p^i (1-p)^{n-i}$$

where α_i is the number of coset leaders of weight i, and p is the symbol error probability.

28. Prove that the binary (23,12,7)-code is perfect.

29. Determine whether or not an (11,6,5)-code over Z_3 is perfect.

30. Consider a linear code with distance $d = 3$. Can any errors of weight 2 be corrected using step-by-step decoding?

31. Let C be a binary (n,k)-code for which the maximum weight of a coset leader in a standard array is w.

(a) Explain how the correspondence table for syndrome decoding can be implemented using an (indexed) array holding only coset leaders.

(b) Show that coset leaders can be stored using at most $m = \min\{n, w\cdot\lceil \log_2 n\rceil\}$ bits. Hence deduce that the storage requirement for syndrome decoding can be reduced to $m2^{n-k}$ bits.

(c) Suppose $n = 70$ and $k = 50$. Using the above results, determine the storage requirement in bits for data structures associated with the syndrome decoding algorithm, assuming (i) $w = 12$, and (ii) $w = 8$.

32. Let C be a binary (n,k)-code for which the maximum weight of a coset leader in a standard array is w. Show that the storage requirement of the correspondence table for step-by-step decoding can be reduced to $\lceil \log_2 w\rceil \cdot 2^{n-k}$ bits.

33. Discuss how the results of the previous two problems change if C is an (n,k)-code over $GF(q)$.

34. Prove or disprove the following statement. In a standard array for a linear code C with distance d, every coset leader has weight at most $t+1$, where $t = \lfloor (d-1)/2 \rfloor$. (cf. exercise 25)

35. Suppose that we want to construct a binary linear code with block length 7 which will correct the error patterns

0000000	0001111
0000001	0011111
0000011	0111111
0000111	1111111

Determine the largest value of k such that a $(7,k)$-code of this type exists.

36. Let C be a $(q+1,2)$-code over $GF(q)$ having minimum distance q.

(a) Determine whether or not C is perfect.

(b) Prove that there is exactly one codeword of C which contains $\gamma \in GF(q)$ in positions i and j for any choice of i and j, $i \neq j$.

(c) Prove that each non-zero codeword contains precisely one zero coordinate.

37. Let the generator matrix for an (n,k)-code C over $GF(3)$ be

$$G = \begin{bmatrix} 2 & 0 & 0 & 2 & 2 & 2 & 1 & 0 \\ 1 & 0 & 2 & 2 & 1 & 2 & 0 & 0 \\ 2 & 1 & 2 & 1 & 2 & 1 & 2 & 1 \\ 0 & 1 & 2 & 1 & 1 & 2 & 2 & 2 \end{bmatrix}.$$

(a) Determine n, k and the number of codewords M in C.

(b) Find a parity check matrix H, and determine the distance d of C.

(c) Find the error pattern associated with the vector $\mathbf{r} = (2\,1\,2\,2\,0\,0\,2\,2)$.

(d) Determine the number of coset leaders in a standard array for C.

(e) Is every vector of weight 2 a coset leader? Explain.

38. Let $H = [I_5\, B]$ be the parity-check matrix for a binary linear code C, where

$$B = \begin{bmatrix} 1 & 0 & 0 & 0 & 0 & 1 \\ 1 & 0 & 1 & 1 & 1 & 1 \\ 1 & 1 & 1 & 0 & 1 & 1 \\ 0 & 1 & 0 & 1 & 1 & 1 \\ 0 & 1 & 1 & 1 & 0 & 1 \end{bmatrix}.$$

(a) Prove that all the codewords in C have even weight.

(b) Determine the distance d of C.

(c) Assuming the first 5 components of a codeword are the check symbols and the last 6 the information symbols, determine the codeword which contains the information symbols 110011.

(d) Assuming that at most one error occurs in transmission, list the 6 information bits corresponding to each of the following received words.
(i) $\mathbf{v}_1 = (11100\ 101000)$ (ii) $\mathbf{v}_2 = (11011\ 110000)$

(e) If $\mathbf{v} = (01110\ 001111)$ is received at the decoder, what can we conclude? Explain.

39. Let the parity check matrix for an (n,k)-code C over $GF(3)$ be

$$H = \begin{bmatrix} 1 & 1 & 1 & 1 & 1 & 1 & 1 & 1 \\ 0 & 1 & 0 & 0 & 1 & 2 & 1 & 2 \\ 0 & 0 & 1 & 0 & 1 & 1 & 2 & 0 \\ 0 & 0 & 0 & 1 & 1 & 1 & 0 & 1 \end{bmatrix}.$$

(a) Determine n, k and the number of codewords M in C.

(b) Find a generator matrix G for C.

(c) Determine the distance d of C.

(d) For each of the following received vectors, indicate the number of channel errors introduced and where possible, correct each to the nearest codeword.
(i) $\mathbf{r}_1 = (0022\,0211)$ (ii) $\mathbf{r}_2 = (0002\,1110)$ (iii) $\mathbf{r}_3 = (1002\,1122)$.

40. $H = [B^T\, I_4]$ is a parity check matrix for a binary (12,4,6)-code C where

$$B = \begin{bmatrix} 1&1&0&0&1&1&0&1 \\ 0&1&1&1&1&1&0&0 \\ 0&0&0&1&1&1&1&1 \\ 1&1&1&1&1&0&1&1 \end{bmatrix}.$$

By the *weight of a coset* we mean the minimum weight of all vectors in the coset.

(a) Determine the number of cosets in a standard array for C.

(b) Determine the number of cosets of weight 2.

(c) Can a coset of C contain both a vector of weight 2 and one of weight 3?

(d) Using the partial list of coset leaders and associated syndromes given below, decode each of the following vectors:
(i) (1100 1000 0001)
(ii) (1101 1110 0100)

Coset Leader		Syndrome
(1010 1000 0000)	$\longleftrightarrow$	(0101 0010)
(1000 0100 1000)	$\longleftrightarrow$	(1000 0101)
(1110 0000 0000)	$\longleftrightarrow$	(1010 1110)
(0110 0010 0000)	$\longleftrightarrow$	(0100 0011)

41. Let C be an (n,k,d) binary code such that there is at least one codeword in C of odd weight. Let S be the set of odd weight codewords in C. Show that $C' = C \backslash S$ is an $(n,k-1,d')$ code with $d' > d$ if d is odd. (This technique is called *expurgating* a code.)

42. Let C be the code given in problem 41 above. Show by example that it is possible that $d' > d+1$.

43. Let the generator matrix for an (n,k)-code C over Z_3 be

$$G = \begin{bmatrix} 1&0&2&0&1&2&1&2 \\ 0&2&2&1&0&2&0&1 \\ 1&0&2&2&1&2&0&0 \\ 0&2&2&1&1&0&0&2 \end{bmatrix}.$$

(a) Determine n and k, and find a parity check matrix H for C.

(b) Determine the distance d of C.

(c) Find the error pattern associated with the vector $\mathbf{r} = (0022\ 0122)$.

44. Let C_1 be a binary (n,k)-code having distance d_1 and generator matrix $G_1 = [I_k\ A]$. Let C_2 be a binary (m,l)-code having distance d_2 with generator matrix $G_2 = [I_l\ B]$. For any $l \times k$ binary matrix K form the $m \times n$ matrix $U_K = \begin{bmatrix} K & S \\ T & R \end{bmatrix}$, where $S = KA$, $T = B^T K$ and $R = B^T KA$. Let $V = \{U_K : \forall K\}$.

(a) Prove that V is a binary (nm, kl)-code where vector addition is matrix addition.

(b) Prove that V has distance $d_1 d_2$.

45. Let C_1 be a binary (n,k,d_1)-code with generator matrix $G_1 = [A\ I_k]$ and let C_2 be a binary (m,k,d_2)-code with generator matrix $G_2 = [B\ I_k]$. Show that

$$H = \begin{bmatrix} & A^T \\ I_{n+m-k} & I_k \\ & B^T \end{bmatrix}$$

is a parity-check matrix for a binary $(n+m, k, d)$-code C where $d \geq d_1 + d_2$.

46. Let C_1 be a binary (n_1,k_1,d_1) code with generator matrix G_1, and let C_2 be a binary (n_2,k_2,d_2) code with generator matrix G_2. Let C be the binary (n,k,d) code with generator $G = \begin{bmatrix} G_1 & 0 \\ 0 & G_2 \end{bmatrix}$. Determine n, k and d.

47. Let C be a binary (n,k)-code whose generator matrix G contains no zero column. Arrange all of the codewords of C as rows of a $2^k \times n$ matrix A.

(a) Show that each column of A contains 2^{k-1} zeros and 2^{k-1} ones.

(b) Using (a), prove that C has distance $d \le (2^{k-1}n)/(2^k-1)$.

(c) Is it possible to find a binary (15,7)-code with minimum distance greater than 7? Explain.

48. Construct a parity check matrix for a (10,6)-code C over $GF(11)$ which has distance at least 5. How many codewords does C have? (Hint: recall example 13)

49. Let S be a k-dimensional subspace of $V_n(F)$ where $F = GF(q)$. Determine the number of t-dimensional subspaces of $V_n(F)$ which contain S, over all $t \ge k$.

50. Explain how to alter the standard array decoding scheme of §3.6 so that for a t-error-correcting linear code, all errors of weight t are corrected and some errors of weight $t+1$ are detected.

51. Prove that for a group G with subgroup H, congruence modulo H is an equivalence relation on the elements of G (cf. exercise 13, Chapter 2).

52. Let C be a linear code with distance d. Prove that every coset of C is a set of vectors having distance d.

53. Generalize the step-by-step decoding technique to non-binary fields.

54. Let C be an (n,k)-code over a field F with distance d. Suppose codewords are transmitted over a channel which introduces *erasures* rather than errors. A component is either received as transmitted (without error), or received as a blank (erased by the channel). For example, for $F = Z_5$, a codeword (01234) might be received as (0_23_) if two erasures occurred. Note that the position of an erased component is known.

(a) Let u be the maximum number of erasures that C can decode in general. Determine u as a function of d. Justify your answer by proving that codewords suffering u erasures are decodable, but those with $u+1$ erasures are not.

(b) Outline a technique for decoding a received vector which has suffered u or fewer erasures.

55. Suppose the codewords of a binary (n,k)-code C are transmitted over a channel which has the property that only the first s coordinate positions of any codeword are subject to error (for some fixed $s<n$), but all error patterns of weight s or less within these positions are equally probable.

(a) Prove that a necessary and sufficient condition for C to be able to correct these error patterns is that all such patterns occur in distinct cosets of a standard array for C.

(b) Describe how to modify the syndrome decoding (i.e. revised standard array decoding) technique to properly decode vectors from a code C satisfying this condition, which are transmitted over this channel.

(c) Does a decoding technique that corrects errors over this channel follow the nearest neighbour decoding strategy? Explain.

56. (a) Construct a parity-check matrix for a (7,4)-code C over $GF(3)$ which is single error correcting.

(b) Determine the number of codewords in C.

(c) Find a generator matrix G for C.

57. In the decoding algorithm of §3.5, why will step (5) never be encountered when decoding a Hamming code?

58. Let C be a linear code having C_i as one of its cosets. Prove that if the coset leader $\mathbf{l}$ of C_i has weight w, then $d(\mathbf{x}, \mathbf{c}) \geq w$ for all $\mathbf{x} \in C_i$ and all $\mathbf{c} \in C$.

59. Prove that using step-by-step decoding on binary codes will correct precisely those errors which are coset leaders in the lexicographic standard array. (Hint: see exercise 58)

60. Prove that

$$H = \begin{bmatrix} 0 & 1 & 0 & 1 & 0 & 1 & 0 & 1 \\ 1 & 0 & 1 & 0 & 0 & 1 & 0 & 1 \\ 0 & 1 & 1 & 0 & 1 & 0 & 0 & 1 \\ 0 & 1 & 0 & 1 & 0 & 0 & 1 & 1 & 0 \end{bmatrix}$$

is the parity-check matrix for a binary linear code of distance 4. (Hint: prove the more general statement that a binary code whose parity-check matrix has columns all of which are nonzero, distinct and of odd weight, has distance at least 4)

61. Suppose that we want to construct a binary linear code with block length 8 and distance 3 which will correct the error patterns:

1 1 0 0 0 0 0 0   1 0 0 0 0 1 0 0
1 0 1 0 0 0 0 0   1 0 0 0 0 0 1 0
1 0 0 1 0 0 0 0   1 0 0 0 0 0 0 1
1 0 0 0 1 0 0 0

Determine the largest value of k such that an $(8,k)$-code of this type exists.

62. (This generalizes exercise 61.) Let C be a binary linear code having block length n and distance 3 which will correct the error patterns:

$$\begin{array}{c} 1\,1\,0\,0 \cdots 0\,0 \\ 1\,0\,1\,0 \cdots 0\,0 \\ 1\,0\,0\,1 \cdots 0\,0 \\ \cdot \\ \cdot \\ \cdot \\ 1\,0\,0\,0 \cdots 1\,0 \\ 1\,0\,0\,0 \cdots 0\,1 \end{array}$$

Determine the largest value of k for which an (n,k)-code of this type exists.

63. Consider a binary (7,3)-code C with parity check matrix

$$H = \left[I_4 \begin{array}{ccc} 1 & 1 & 0 \\ 0 & 1 & 1 \\ 1 & 0 & 1 \\ 1 & 1 & 0 \end{array} \right]$$

(a) Construct a 1-1 correspondence between syndromes and weights of the lightest vectors in the corresponding cosets.

(b) Decode each of the following by step-by-step decoding:

$$\mathbf{r}_1 = (1\,1\,1\,1\,1\,0\,1)$$

$$\mathbf{r}_2 = (1\,1\,0\,1\,0\,1\,1).$$

64. Prove or disprove: There exists a binary $(2^m, 2^m - m)$ single error correcting, double error detecting code, $m \geq 3$.

65. Suppose C is a linear code over $GF(q^m)$, $m \geq 2$. Let S be the set of codewords such that all components of these codewords are in $GF(q)$. Show that S is also a linear code over $GF(q)$. Is S a linear code over $GF(q^m)$?

66. A *repetition code* is a code such that the parity check symbols of any codeword simply repeat the information symbols (i.e. all codewords take the form $(x_1, x_2, \ldots, x_k, x_1, x_2, \ldots, x_k, \ldots, x_1, x_2, \ldots, x_k)$.).

i) How many codewords are there in a repetition code with k information symbols over an alphabet A?

ii) Construct the parity check matrix for such a code.

67. Suppose $\mathbf{x} = (x_1,x_2,\ldots,x_n)$, $\mathbf{y} = (y_1,y_2,\ldots,y_n) \in V_n(Z_3)$. Show that $w(\mathbf{x}+\mathbf{y}) = w(\mathbf{x})+w(\mathbf{y})-2c$ where c is the number of 2's in $(x_1y_1,x_2y_2,\ldots,x_ny_n)$.

Bounds for Codes

68. (a) Prove that a t-error-correcting (n,k)-code over $GF(q)$ satisfies the *Hamming bound:* $q^n \geq q^k \sum_{i=0}^{t} \binom{n}{i}(q-1)^i$. (Hint: consider the spheres of radius t about each codeword)

 (b) Determine whether or not a binary (12,7,5)-code C exists. Explain your reasoning.

69. (a) Prove that in an (n,k,d)-code C over a field F, $d \leq n-k+1$. This is known as the *Singleton bound.*

 (b) Prove that an (n,k)-code C capable of correcting t errors must have at least $2t$ check symbols.

70. Let C be an (n,k,d)-code over a field F where $d = n-k+1$. Such a code is called a *maximum distance separable (MDS)* code.

 (a) Construct two examples of MDS codes.

 (b) Which of the Hamming codes are MDS codes?

71. Prove that there exists an (n,k)-code C over $GF(q)$ having distance d provided that $\sum_{i=0}^{d-2}(q-1)^i\binom{n-1}{i} < q^{n-k}$. (This is the *Gilbert-Varshamov bound* for q-ary codes.)

For problems 72–76 let C be an (n,k)-binary code. This sequence of problems leads to the derivation of what is called the Elias upper bound *for binary codes.*

72. Let M_i denote the number of codewords of C in a sphere of radius t having the ith n-tuple as its centre, $1 \leq i \leq 2^n$. Show that $\sum_{i=1}^{2^n} M_i = 2^k \sum_{j=0}^{t} \binom{n}{j}$.

73. Prove that there exists a sphere S_M of radius t around an n-tuple $\mathbf{x}$ that contains M codewords where $M \geq (\sum_{j=0}^{t}\binom{n}{j})2^{k-n}$.

74. Let A be an array whose rows are the vectors $\mathbf{c}_i - \mathbf{x}$ where $\mathbf{c}_i \in S_M \cap C$, $1 \leq i \leq M$, and $\mathbf{x}$ and M are given by problem 75. Let α_j be the number of rows in column j of A. Show that $\sum_{j=1}^{n}\alpha_j = Mt-\Delta$ and $\sum_{j=1}^{n}\alpha_j^2 \geq (Mt-\Delta)^2/n$, for some $\Delta \geq 0$.

75. Let d_T denote the sum of the $\binom{M}{2}$ distances between the rows of A. Show that $d_T \le tM^2(1-t/n)$.

76. Let $d_{\min}$ be the minimum distance in $S_M \cap C$. Prove that the distance d in C satisfies

$$d \le d_{\min} \le 2t(1-\frac{t}{n})(\frac{M}{M-1}).$$

Problems 77–80 are all concerned with what is called the Plotkin bound *for codes.*

77. Let C be a binary (n,k)-code. Let d be the distance of C. Show that $d \le n2^{k-1}/(2^k-1)$. (Hint: See problem 18(a) and consider a lower bound for the sum of all distances between pairs of distinct codewords.) This is the *Plotkin bound* for binary (n,k)-codes.

78. Let C be a binary $[n,M,d]$-code.

    (i) Let s denote the sum of the distances between pairs of distinct codewords in C. Show that $s \ge \binom{M}{2}d$.

    (ii) Arrange the codewords of C as rows of a matrix A. Let x_i denote the number of 1's in column i of A. Show that $s = \sum_{i=1}^{n} x_i(M-x_i)$.

    (iii) Prove that

$$d \le \frac{nM}{2(M-1)}.$$

    (This is the *Plotkin bound* for binary [n,M,d] codes).

79. Let C be an (n,k)-code over $GF(q)$ having distance d. Prove the Plotkin bound

$$d \le n\,\frac{q^{k-1}(q-1)}{(q^k-1)}.$$

    (This generalizes problem 77.)

80. Let C be an $[n,M,d]$-code over $GF(q)$ having distance d. Prove the Plotkin bound for C:

$$d \le n(q-1)M/q(M-1).$$

    (This generalizes problem 78.)

Weight Distribution and Enumerators

81. Let C be an $[n,M]$-code and let $(A_0,A_1,\cdots,A_n)$ be the weight distribution of C. If $\mathbf{x} \in C$ let d_i denote the number of n-tuples at distance i from $\mathbf{x}$. Prove that $(d_0,d_1,\cdots,d_n)$ is equal to the weight distribution of C for all $\mathbf{x} \in C$ if C is linear. Is the result true if C is not linear?

82. Let C be a binary code with distance $d \geq 3$ and block length n with weight distribution $(A_0,A_1,\ldots,A_n)$.

    (i) Show that there are exactly iA_i binary n-tuples of weight $i-1$ at distance 1 from codewords of C.

    (ii) Show that there are exactly $(n-i)A_i$ binary n-tuples of weight $i+1$ at distance 1 from codewords of C.

83. Show that for a binary Hamming code the weight distribution satisfies the recurrence $A_0=1, A_1=0,$

$$(i+1)A_{i+1}+A_i+(n-i+1)A_{i-1}=\binom{n}{i}\,,\ i\geq 1\,,$$

and, hence, that

$$\sum_{i=1}^{n} iA_ix^{i-1}+\sum_{i=0}^{n} A_ix^i+\sum_{i=0}^{n}(n-i)A_ix^{i+1}=(1+x)^n\,.$$

84. Let C be an (n,k)-code over $GF(q)$ with weight distribution $(A_0,A_1,\cdots,A_n)$ and let $(A_0',A_1',\cdots,A_n')$ be the weight distribution of $C^{\perp}$. Prove that

$$\sum_{i=0}^{n-j}\binom{n-i}{j}A_i=q^{k-j}\sum_{i=0}^{j}\binom{n-i}{j-i}A_i',\ \ j=0,1,\ldots,n\,.$$

85. Let C be a binary (n,k)-code which contains the all ones vector. If $(A_0,A_1,\cdots,A_n)$ is the weight distribution for C then prove that $A_i=A_{n-i}$ for $0 \leq i \leq n$.

86. Let C be an (n,k)-code over $GF(q)$ with weight distribution $(A_0,A_1,\cdots,A_n)$. Define

$$W_C'(z)=\sum_{i=0}^{n} A_iz^i\,.$$

Prove that

$$W'_{C^\perp}(z) = \frac{1}{q^k}[1+(q-1)z]^n W'_C\left(\frac{1-z}{1+(q-1)z}\right).$$

(Hint. This is easily deduced from Theorem 3.13).

87. Let C b a binary (10,7)-code with parity check matrix

$$H = \begin{bmatrix} 1 & 0 & 0 & 1 & 0 & 1 & 1 & 1 & 1 & 1 \\ 0 & 1 & 0 & 1 & 1 & 1 & 1 & 0 & 1 & 1 \\ 0 & 0 & 1 & 0 & 1 & 1 & 0 & 1 & 1 & 0 \end{bmatrix}.$$

Determine the weight enumerator for C.

88. Let C be a binary linear code and let E be the set of all even weight codewords. Recall that E is itself a code. Using the definition in exercise 86, prove that

$$W'_E(z) = \frac{1}{2}[W'_C(z) + W'_C(z)].$$

89. Let C be a binary linear code and let C' be the code obtained from C by adding an overall parity bit to each codeword (see problem 24). Using the definition in exercise 86, prove that

$$W'_{C'}(z) = \frac{1}{2}[(1+z)W'_C(z) + (1-z)W'_C(-z)].$$

90. Prove that $a(x,y) = x^5 + 3x^3y^2 + 9x^2y^3 + 14xy^4 + y^5$ cannot be the weight enumerator for a linear code over $GF(3)$.

91. Prove that $f(x,y) = x^5 + 2x^4y + 2x^2y^3 + 2xy^4 + y^5$ is not the weight enumerator for a binary linear code.

92. Prove that there is no (6,4)-code over Z_3 having weight distribution (1,0,5,18,36,12,9).

93. Construct a binary (5,3)-code C having weight enumerator

$$W_C(x,y)\$\$= x^5 + x^4y + 2x^3y^2 + 2x^2y^3 + xy^4 + y^5.$$

(Hint: Construct the dual code $C^\perp$ first.)

94. Construct a (6,4)-code over Z_3 having weight distribution (1,0,6,16,36,12,10).

95. Does there exist a (6,4)-code over $GF(3)$ having weight distribution (1,0,4,18,36,14,8)?

96. Let $W_C(x,y)$ be the weight enumerator for a binary linear code. Prove that $W_C(1,-1)$ equals either 0 or $|C|$.

97. Let C be a linear code over $GF(q)$ having weight distribution $(A_0,A_1,\dots,A_n)$. Prove that $q-1$ divides A_i, $1 \le i \le n$.

98. Let $\mathbf{c}$ be a codeword in a linear code of block length n over $GF(q)$, and suppose $w(\mathbf{c}) = i$. As in §3.8, let $N(i,h;s)$ denote the number of n-tuples over $GF(q)$ of weight h which are distance s from $\mathbf{c}$. Show that

$$N(i,h;s) = \sum_{\substack{0 \le j,k,l \le n \\ i+j-l=h \\ j+k+l=s}} \left[\binom{n-i}{j}(q-1)^j\right]\left[\binom{i}{k}(q-2)^k\right]\binom{i-k}{l}$$

and hence that $N(i,h;s)$ is independent of the particular codeword $\mathbf{c}$ of weight i. (Hint: Consider changing j of the zero components of $\mathbf{c}$ to nonzero elements, changing k of the nonzero components to other nonzero elements, and changing l of the other nonzero elements to zero). See also Chapter 1, exercise 30.

Chapter 4

SOME SPECIAL LINEAR CODES

4.1. Introduction

In this chapter we describe two special classes of linear codes. One reason for our interest in them is the fact that they both have been used in space exploration projects. The first class we consider was used to transmit photographs from the *Mariner 9* spacecraft in January of 1972. The second class was used to transmit colour photographs of Jupiter in March and July of 1979, and of Saturn in November 1980 and August 1981 from the *Voyager* 1 and *Voyager* 2 spacecraft.

4.2 First Order Reed-Muller Codes†

Let H_r be the parity-check matrix for the $(2^r-1, 2^r-1-r)$-binary Hamming code, as discussed in §3.4. Its $2^r - 1$ columns consist of all possible non-zero r-tuples. Define $B_r = [H_r\ 0]$, i.e. H_r with a column of 0's adjoined. Let $\mathbf{v}_1, \mathbf{v}_2, \ldots, \mathbf{v}_r$ be the rows of B_r and let $\mathbf{1}$ be the row vector of length 2^r consisting of all 1's.

Definition. The *first-order Reed-Muller code*, denoted $R(1,r)$, is the subspace generated by the vectors $\mathbf{1}, \mathbf{v}_1, \mathbf{v}_2, \ldots, \mathbf{v}_r$. (For a definition of and some results on kth-order Reed-Muller codes the reader is referred to exercises 28 through 42.)

$R(1,r)$ is a binary code with generator matrix

$$G = \left[\frac{\mathbf{1}}{B_r}\right] = \left[\frac{\mathbf{1}}{H_r\ 0}\right].$$

We begin by establishing the parameters of $R(1,r)$. It turns out that the distance of the code is half the length of the codewords.

Theorem 4.1. $R(1,r)$ is a $(2^r, r+1, 2^{r-1})$-code.

† This section may be omitted without loss of continuity.

Proof.

Let the parameters of the code be (n,k,d). Since the generator G has 2^r columns, it is clear that $n = 2^r$. Note that G has $r+1$ rows. In fact, these rows are linearly independent, because the $r \times r$ identity matrix can be found within the submatrix B_r, establishing that the last r rows of G are linearly independent, and certainly the first row is independent from all others. Hence $k = r+1$, and $R(1,r)$ is a $(2^r,r+1)$-code. It remains only to show that the distance is $d = 2^{r-1}$. We do this by showing that every non-zero codeword has weight 2^{r-1}, except for the all-ones codeword $\mathbf{1}$. It then follows that the minimum non-zero weight is 2^{r-1} and the distance is the same.

Any non-zero codeword $\mathbf{c}$ has the form

$$\mathbf{c} = \mathbf{r}_{i_1} + \mathbf{r}_{i_2} + \cdots + \mathbf{r}_{i_h}$$

where $\mathbf{r}_{i_j}$ is the i_j^{th} row of G, $1 \le h \le r+1$. Suppose no $\mathbf{r}_{i_j}$ is $\mathbf{1}$. Consider the matrix

$$A = \begin{bmatrix} \mathbf{r}_{i_1} \\ \mathbf{r}_{i_2} \\ \cdot \\ \cdot \\ \mathbf{r}_{i_h} \end{bmatrix}.$$

Now component t of $\mathbf{c}$ will be 0 if the tth column of A has an even number of 1's, and 1 if that column of A contains an odd number of 1's. Note that for each distinct column $\hat{\mathbf{u}}$ of A, the number of columns $\mathbf{u}$ of B_r whose i_j^{th} component, $1 \le j \le h$, agrees with component j of $\hat{\mathbf{u}}$ is 2^{r-h}, since each of the remaining $r-h$ components of $\mathbf{u}$ are free to take on 0 or 1 (and by definition every distinct binary r-tuple appears as a column of B_r). Hence each distinct column of A appears precisely 2^{r-h} times in A. But every distinct binary h-tuple appears as a column of A, and since there are as many even-weight binary h-tuples as there are odd-weight binary h-tuples, it follows that exactly half the components of $\mathbf{c}$ are 1's, and thus $w(\mathbf{c}) = 2^{r-1}$ (cf. exercise 19, Chapter 3).

If some $\mathbf{r}_{i_j} = \mathbf{1}$, then consider $\mathbf{c} - \mathbf{1}$. The above argument, which was valid for any non-zero codeword, then gives $w(\mathbf{c}-\mathbf{1}) = 2^{r-1}$, as long as $\mathbf{c} \ne \mathbf{1}$. It then follows that $w(\mathbf{c}) = 2^{r-1}$ by changing 1's and 0's in $\mathbf{c} - \mathbf{1}$. This completes the proof. □

We now give an example of a first order Reed-Muller code $R(1,r)$ for $r = 3$.

Example 1.

$$H_3 = \begin{bmatrix} 1 & 0 & 0 & 1 & 0 & 1 & 1 \\ 0 & 1 & 0 & 1 & 1 & 0 & 1 \\ 0 & 0 & 1 & 0 & 1 & 1 & 1 \end{bmatrix}$$

$$B_3 = [H_3\ \mathbf{0}] = \begin{bmatrix} 1 & 0 & 0 & 1 & 0 & 1 & 1 & 0 \\ 0 & 1 & 0 & 1 & 1 & 0 & 1 & 0 \\ 0 & 0 & 1 & 0 & 1 & 1 & 1 & 0 \end{bmatrix}$$

$$G = \left[\frac{\mathbf{1}}{B_r}\right] = \left[\frac{\mathbf{1}}{H_r\ \mathbf{0}}\right] = \begin{bmatrix} 1 & 1 & 1 & 1 & 1 & 1 & 1 & 1 \\ 1 & 0 & 0 & 1 & 0 & 1 & 1 & 0 \\ 0 & 1 & 0 & 1 & 1 & 0 & 1 & 0 \\ 0 & 0 & 1 & 0 & 1 & 1 & 1 & 0 \end{bmatrix}.$$

G generates $R(1,3)$. All vectors in the code have weight 4 except $\mathbf{0}$ and $\mathbf{1}$. $R(1,3)$ is an (8,4,4)-code.

For the code $R(1,r)$, to decode a received vector $\mathbf{r}$ we shall make use of what is called the *Hadamard transform* of a vector $\mathbf{R}$ derived from $\mathbf{r}$. The Hadamard transform is related to a special type of matrix, called a *Hadamard matrix*. Before presenting the Hadamard transform and a decoding procedure for $R(1,r)$, we briefly introduce the Hadamard matrix.

Definition. A *Hadamard matrix* H_n of order n is an $n \times n$ matrix with integer entries +1 and −1, whose rows are pairwise orthogonal as real numbers.

By "pairwise orthogonal as real numbers" we mean that the inner product over the real numbers (as opposed to over another field such as Z_2) of distinct rows is 0. Regarding notation, the context should rule out any possible confusion between the Hadamard matrix H_n and the parity-check matrix H_r for the Hamming code of order r discussed earlier.

Example 2.

Hadamard matrices of orders 1, 2 and 4 are given by

$$H_1 = [1] \qquad H_2 = \begin{bmatrix} 1 & 1 \\ 1 & -1 \end{bmatrix} \qquad H_4 = \begin{bmatrix} 1 & 1 & 1 & 1 \\ 1 & -1 & 1 & -1 \\ 1 & 1 & -1 & -1 \\ 1 & -1 & -1 & 1 \end{bmatrix}.$$

An equivalent definition to the one given above is that H_n is an $n \times n$ matrix with integer entries $+1$ and -1 such that

$$H_n H_n^T = nI_n.$$

The following properties of H_n are easily established.

(1) $H_n^{T'} = nH_n^{-1}$ and so $H_n^T H_n = nI_n$. It follows that whenever H_n is a Hadamard matrix, so is H_n^T, and that the columns of H_n are orthogonal as well as the rows.

(2) Exchanging rows or columns changes one Hadamard matrix into another one.

(3) Multiplying rows or columns by -1 changes one Hadamard matrix into another.

Given a Hadamard matrix H_n of order n, we can construct the matrix

$$H_{2n} = \begin{bmatrix} H_n & H_n \\ H_n & -H_n \end{bmatrix}.$$

It is easy to verify that H_{2n} is in fact a Hadamard matrix of order $2n$. Using this construction and starting with $H_1 = [1]$, we can construct H_2, H_4, and indeed, Hadamard matrices H_{2^r} of order 2^r for all $r \geq 1$. We shall see that these particular Hadamard matrices of order 2^r are of use when we apply the Hadamard transform to Reed-Muller codes.

The above construction can be viewed as a *Kronecker product* of matrices.

Definition. Let $A = [a_{ij}]$ and $B = [b_{ij}]$ be square matrices of orders m and n respectively. The *Kronecker product* of A with B, denoted $A \times B$, is the $mn \times mn$ matrix

$$A \times B = [a_{ij}B]$$

obtained by replacing the (i,j)-entry of A with the matrix $a_{ij}B$, the matrix B with each entry multiplied by a_{ij}.

If we let $n = 2^{r-1}$, then we see that the above construction can alternatively be specified

$$H_{2n} \ (= H_{2^r}) = H_2 \times H_2 \times \cdots \times H_2,$$

the Kronecker product of r matrices H_2 (with H_2 as given in example 2). We reserve the notation $H(2^r)$ for the Hadamard matrix of order 2^r of this particular form.

We now consider a second construction. First, we define the *proper ordering* on binary r-tuples.

Definition. The *proper ordering* P_r of binary r-tuples is the ordering defined recursively by the rules

(1) $P_1 = [0,1]$

(2) if $P_i = [\mathbf{b}_1, \mathbf{b}_2, ..., \mathbf{b}_{2^i}]$ then $P_{i+1} = [\mathbf{b}_1 0, \mathbf{b}_2 0, ..., \mathbf{b}_{2^i} 0, \mathbf{b}_1 1, \mathbf{b}_2 1, ..., \mathbf{b}_{2^i} 1]$,
for $1 \le i \le r-1$.

Example 3.

$P_1 = [0,1]$
$P_2 = [00, 10, 01, 11]$
$P_3 = [000, 100, 010, 110, 001, 101, 011, 111]$.

Now let $n = 2^r$, and let $\mathbf{u}_0, \mathbf{u}_1, ..., \mathbf{u}_{n-1}$ be the binary r-tuples in proper order. If we agree that the leftmost bit of an r-tuple is to be read as the least significant bit, then the components of $\mathbf{u}_i$ are the r-bit representation of the integer i. Consider constructing a matrix H with entries +1 and −1 as follows. Index the rows of H by the integers 0 through $n-1$, and the columns likewise (rather than the standard 1 through n). Let H be defined by $H = [h_{ij}]$ with (i,j)-entry

$$h_{ij} = (-1)^{\mathbf{u}_i \cdot \mathbf{u}_j}.$$

A small example serves to illustrate.

Example 4.

Consider $r = 2$. If we index the rows and columns of a 4×4 matrix H with the binary 2-tuples in the proper order P_2, and construct $H = [h_{ij}]$ with $h_{ij} = (-1)^{\mathbf{u}_i \cdot \mathbf{u}_j}$, then

$$H = \begin{bmatrix} 1 & 1 & 1 & 1 \\ 1 & -1 & 1 & -1 \\ 1 & 1 & -1 & -1 \\ 1 & -1 & -1 & 1 \end{bmatrix}.$$

Note that this matrix H is the same as the matrix $H_2 \times H_2$ where $H_2 = H(2^1)$.

Using induction on r, it is easily established that the matrix H so constructed is always a Hadamard matrix (exercise 1). In fact, it is the matrix $H(2^r)$, and the two constructions produce the same matrix.

Before proceeding to define the Hadamard transform, we introduce the following notation. Let $\mathbf{r}$ be a binary 2^r-tuple. To each component of the vector $\mathbf{r}$ associate a distinct binary r-tuple. Since there are precisely 2^r binary r-tuples, this is possible. For the binary r-tuple $\mathbf{u}$ define the scalar $\mathbf{r}(\mathbf{u})$ to be the component of $\mathbf{r}$ associated with $\mathbf{u}$. From $\mathbf{r}$ we construct a real 2^r-tuple $\mathbf{R}$ where the component of the vector $\mathbf{R}$ associated with $\mathbf{u}$ is the real scalar

$$\mathbf{R}(\mathbf{u}) = (-1)^{\mathbf{r}(\mathbf{u})}.$$

$\mathbf{R}$ is then a vector with real entries ± 1, and since $\mathbf{r}$ is a binary vector, $\mathbf{r}(\mathbf{u})$ is 0 or 1.

Example 5.

For $r = 3$, consider $\mathbf{r} = (1101\ 1100)$. The 8 distinct binary 3-tuples can be found as the columns of B_3 from example 1. Associate the 3-tuple from the ith column of B_3 with the ith component of $\mathbf{r}$. For example, $(110)^T$ is associated with the 4th component of $\mathbf{r}$, a 1. Then $\mathbf{r}(\,(110)\,)$ picks out the component of $\mathbf{r}$ associated with (110), i.e. the 4th component. Hence $\mathbf{r}(\,(110)\,) = 1$. Now from $\mathbf{r}$, we derive $\mathbf{R}$. The component of $\mathbf{R}$ associated with (110) is defined to be

$$\mathbf{R}(\,(110)\,) = (-1)^{\mathbf{r}(110)} = (-1)^1 = -1.$$

Hence the 4th component of $\mathbf{R}$ is -1. Computing the other components, we get

$$\mathbf{R} = (-1, -1, 1, -1, \ -1, -1, 1, 1).$$

We see that to derive **R** from **r**, we simply map 0's and 1's in **r** to 1's and −1's, respectively, in **R**. The preceding example makes use of our notation to express this mapping formally through the association of distinct binary r-tuples with unique components of a 2^r-tuple. This association is a concept necessary for the Hadamard transform. With **R** derived from **r** as outlined above, we can now proceed to define the Hadamard transform. In this section, we shall use the abbreviation V_r for $V_r(Z_2)$, the set of all binary r-tuples.

Definition. The *Hadamard transform* of the 2^r-tuple **R** is the 2^r-tuple $\hat{\mathbf{R}}$ where the component of $\hat{\mathbf{R}}$ associated with the r-tuple $\mathbf{u} \in V_r$ is

$$\hat{\mathbf{R}}(\mathbf{u}) = \sum_{\mathbf{v} \in V_r} (-1)^{\mathbf{u}\cdot\mathbf{v}} \mathbf{R}(\mathbf{v}).$$

With the components of **R** being integers +1 and −1, the components of $\hat{\mathbf{R}}$ will be integers. Since

$$\mathbf{R}(\mathbf{v}) = (-1)^{\mathbf{r}(\mathbf{v})},$$

we have

$$\hat{\mathbf{R}}(\mathbf{u}) = \sum_{\mathbf{v} \in V_r} (-1)^{\mathbf{u}\cdot\mathbf{v}+\mathbf{r}(\mathbf{v})}.$$

Example 6.

For $\mathbf{u} = (110)^T$, using the association of components to vectors as defined by B_3 in example 1, we compute $\hat{\mathbf{R}}((110))$ as follows, for $\mathbf{r} = (1101\ 1100)$.

$$\begin{aligned}\hat{\mathbf{R}}(\,(110)\,) &= \sum_{\mathbf{v} \in V_3} (-1)^{(110)\cdot\mathbf{v}+\mathbf{r}(\mathbf{v})} \\ &= (-1)^{(110)\cdot(100)+\mathbf{r}(100)} + (-1)^{(110)\cdot(010)+\mathbf{r}(010)} \\ &\quad + (-1)^{(110)\cdot(001)+\mathbf{r}(001)} + (-1)^{(110)\cdot(110)+\mathbf{r}(110)} \\ &\quad + (-1)^{(110)\cdot(011)+\mathbf{r}(011)} + (-1)^{(110)\cdot(101)+\mathbf{r}(101)}\end{aligned}$$

$$+(-1)^{(110)\cdot(111)+r(111)}+(-1)^{(110)\cdot(000)+r(000)}$$

$$=\ (-1)^{1+1}+(-1)^{1+1}+(-1)^{0+0}+(-1)^{0+1}$$

$$+(-1)^{1+1}+(-1)^{1+1}+(-1)^{0+0}+(-1)^{0+0}$$

$$=6.$$

Since $\mathbf{u}=(110)^T$ is associated with the 4th component, we have that the 4th component of $\hat{\mathbf{R}}$ is 6. The remaining 7 components could be determined in a similar manner.

An alternate definition of $\hat{\mathbf{R}}$ is

$$\hat{\mathbf{R}}=\mathbf{R}H$$

where $H=[h_{\mathbf{v},\mathbf{u}}]$, $h_{\mathbf{v},\mathbf{u}}=(-1)^{\mathbf{v}\cdot\mathbf{u}}$, $\mathbf{v},\mathbf{u}\in V_r$, is a Hadamard matrix of order 2^r with rows and columns indexed by the 2^r distinct binary vectors of V_r. The two definitions are easily shown to be equivalent (exercise 25). The order of indexing vectors is that the r-tuple associated with component i of $\mathbf{r}$ indexes the ith row and column of H. By convention, we associate r-tuples to coordinate positions using the vectors of V_r in the proper ordering, so that $H=H(2^r)$.

The following theorem provides the basis for a decoding scheme for $R(1,r)$. Let the ordered columns of B_r be used to associate the distinct binary r-tuples with the coordinate positions in $\mathbf{r}$, $\mathbf{R}$ and $\hat{\mathbf{R}}$.

Theorem 4.2. $\hat{\mathbf{R}}(\mathbf{u})$ is the number of 0's minus the number of 1's in the binary vector

$$\mathbf{t}=\mathbf{r}+\sum_{i=1}^{r}u_i\mathbf{v}_i$$

where $\mathbf{u}=(u_1,u_2,\ldots,u_r)^T$ is a binary r-tuple and $\mathbf{v}_i$ is the ith row of B_r.

Proof.

Let $\mathbf{v}_i=(a_{i1},a_{i2},\ldots,a_{in})$, $n=2^r$, $1\le i\le r$. Consider the binary n-tuple

$$\sum_{i=1}^{r}u_i\mathbf{v}_i=\left(\sum_{i=1}^{r}u_ia_{i1},\ \sum_{i=1}^{r}u_ia_{i2},\ldots,\ \sum_{i=1}^{r}u_ia_{in}\right)=(\mathbf{u}\cdot\mathbf{l}_1,\mathbf{u}\cdot\mathbf{l}_2,\ldots,\mathbf{u}\cdot\mathbf{l}_n)=\mathbf{u}B_r,$$

where $\mathbf{l}_j$ is the jth column vector in B_r. Now since we are using the columns of B_r to associate binary r-tuples with components, we have by definition

$$\mathbf{r} = (\, \mathbf{r}(\mathbf{l}_1), \mathbf{r}(\mathbf{l}_2), \ldots, \mathbf{r}(\mathbf{l}_n)\,).$$

Thus

$$\mathbf{t} = \mathbf{r} + \sum_{i=1}^{r} u_i \mathbf{v}_i = (\, \mathbf{u}\cdot\mathbf{l}_1 + \mathbf{r}(\mathbf{l}_1), \mathbf{u}\cdot\mathbf{l}_2 + \mathbf{r}(\mathbf{l}_2), \ldots, \mathbf{u}\cdot\mathbf{l}_n + \mathbf{r}(\mathbf{l}_n)\,).$$

Clearly

$$(-1)^{\mathbf{u}\cdot\mathbf{l}_j + \mathbf{r}(\mathbf{l}_j)} = \begin{cases} -1 & \text{if } \mathbf{u}\cdot\mathbf{l}_j + \mathbf{r}(\mathbf{l}_j) = 1 \\ +1 & \text{if } \mathbf{u}\cdot\mathbf{l}_j + \mathbf{r}(\mathbf{l}_j) = 0 \end{cases}$$

Now from the definition of the Hadamard transform we have that

$$\hat{\mathbf{R}}(\mathbf{u}) = \sum_{\mathbf{v} \in V_r} (-1)^{\mathbf{u}\cdot\mathbf{v} + \mathbf{r}(\mathbf{v})}.$$

Since the columns $\mathbf{l}_j$, $j = 1,\ldots,n$ of B_r are all the distinct r-tuples, i.e. all $\mathbf{v} \in V_r$, this may be written as

$$\hat{\mathbf{R}}(\mathbf{u}) = \sum_{j=1}^{n} (-1)^{\mathbf{u}\cdot\mathbf{l}_j + \mathbf{r}(\mathbf{l}_j)}.$$

Hence $\hat{\mathbf{R}}(\mathbf{u})$ counts +1 for every component in **t** that is 0, and −1 for every component that is 1. □

Using this result, we now develop two equations to be used in a decoding algorithm for $R(1,r)$. We may compute the number of 0's in **t** as

$$t_0 = 2^r - w(\mathbf{t}) = 2^r - w(\,\mathbf{r} + \sum_{i=1}^{r} u_i \mathbf{v}_i\,) = 2^r - d(\mathbf{r}, \sum_{i=1}^{r} u_i \mathbf{v}_i).$$

From this, the number of 1's in **t** is

$$t_1 = d(\mathbf{r}, \sum_{i=1}^{r} u_i \mathbf{v}_i).$$

Now $\hat{\mathbf{R}}(\mathbf{u}) = t_0 - t_1$ and so

$$\hat{\mathbf{R}}(\mathbf{u}) = 2^r - d(\mathbf{r}, \sum_{i=1}^{r} u_i \mathbf{v}_i) - d(\mathbf{r}, \sum_{i=1}^{r} u_i \mathbf{v}_i) = 2^r - 2d(\mathbf{r}, \sum_{i=1}^{r} u_i \mathbf{v}_i). \quad \text{(A)}$$

Another way to determine $\hat{\mathbf{R}}(\mathbf{u})$ is as follows. The number of 0's in **t** may also be computed as

$$t_0 = w(1+t) = w(1+\mathbf{r}+\sum_{i=1}^{r} u_i \mathbf{v}_i) = d(\mathbf{r}, 1+\sum_{i=1}^{r} u_i \mathbf{v}_i).$$

The number of 1's in **t** is then given by

$$t_1 = 2^r - d(\mathbf{r}, 1+\sum_{i=1}^{r} u_i \mathbf{v}_i).$$

Then since $\hat{R}(\mathbf{u}) = t_0 - t_1$,

$$\hat{R}(\mathbf{u}) = 2d(\mathbf{r}, 1+\sum_{i=1}^{r} u_i \mathbf{v}_i) - 2^r. \tag{B}$$

Rewriting (A) and (B) we obtain the equations

$$d(\mathbf{r}, \sum_{i=1}^{r} u_i \mathbf{v}_i) = \tfrac{1}{2}(2^r - \hat{R}(\mathbf{u}))$$

and (C)

$$d(\mathbf{r}, 1+\sum_{i=1}^{r} u_i \mathbf{v}_i) = \tfrac{1}{2}(2^r + \hat{R}(\mathbf{u}))$$

We now discuss how to use these equations in a decoding scheme for $R(1,r)$. Recall that the $\mathbf{v}_i$ are rows of B_r, and

$$G = \left[\frac{\mathbf{1}}{B_r}\right]$$

is a generator for $R(1,r)$. Now for any binary r-tuple $\mathbf{u} = (u_1, ..., u_r)$,

$$\mathbf{u}B_r = \sum_{i=1}^{r} u_i \mathbf{v}_i.$$

Suppose **r** is a received vector. We wish to decode **r** to the codeword closest to **r**. Message $(r+1)$-tuples look like $\mathbf{m} = (u_0,\mathbf{u})$ where **u** is some r-tuple and $u_0 = 0$ or 1. The corresponding transmitted codeword is

$$\mathbf{c} = \mathbf{m}G = (u_0,\mathbf{u})\left[\frac{\mathbf{1}}{B_r}\right] = u_0 \cdot \mathbf{1} + \sum_{i=1}^{r} u_i \mathbf{v}_i.$$

The two equations in (C) consider the two cases $u_0 = 0$ and $u_0 = 1$, respectively. To find the codeword closest to **r**, we find the codeword **c** minimizing $d(\mathbf{r},\mathbf{c})$. Note that (C) gives us precisely $d(\mathbf{r},\mathbf{c})$ (as a function of $\hat{R}(\mathbf{u})$) for $\mathbf{c} = \sum_{i=1}^{r} u_i \mathbf{v}_i$ and $\mathbf{c} = \mathbf{1} + \sum_{i=1}^{r} u_i \mathbf{v}_i$. Computing $\hat{R}(\mathbf{u})$ over all $\mathbf{u} \in V_r$, together with these two cases, is hence equivalent to computing the distance from $\mathbf{r}$ to every codeword. From (C) we see that this distance is minimized by $\mathbf{u}$ which minimizes

$$\min \{ 2^r - \hat{\mathbf{R}}(\mathbf{u}), 2^r + \hat{\mathbf{R}}(\mathbf{u}) \},$$

i.e., by $\mathbf{u}$ which maximizes the magnitude of $\hat{\mathbf{R}}(\mathbf{u})$. Suppose we find a $\mathbf{u}$ maximizing $|\hat{\mathbf{R}}(\mathbf{u})|$, and this maximum is unique over all $\mathbf{u} \in V_r$. Then there is a unique codeword $\mathbf{c}$ closest to $\mathbf{r}$. For such a vector $\mathbf{u}$, if $\hat{\mathbf{R}}(\mathbf{u}) > 0$, the minimum distance occurs in the first equation of (C). If $\hat{\mathbf{R}}(\mathbf{u}) < 0$, the minimum distance occurs in the second equation. A decoding scheme for the first order Reed-Muller code now follows.

Decoding Algorithm for first-order Reed-Muller codes.

Let $\mathbf{r}$ be a received binary vector of length 2^r.

Let the columns of B_r be in the proper ordering P_r.

Let H be the Hadamard matrix $H = H(2^r)$.

(1) Compute $\mathbf{R}$ and $\hat{\mathbf{R}}$, where $\mathbf{R}(\mathbf{u}) = (-1)^{\mathbf{r}(\mathbf{u})}$ and $\hat{\mathbf{R}} = \mathbf{R}H$.

(2) Find a component $\hat{\mathbf{R}}(\mathbf{u})$ of $\hat{\mathbf{R}}$ whose magnitude is maximum; let $\mathbf{u} = (u_1, \dots, u_r)^T$.

(3) If $\hat{\mathbf{R}}(\mathbf{u}) > 0$, then decode $\mathbf{r}$ as $\sum_{i=1}^{r} u_i \mathbf{v}_i$.

(4) If $\hat{\mathbf{R}}(\mathbf{u}) \le 0$, then decode $\mathbf{r}$ as $\mathbf{1} + \sum_{i=1}^{r} u_i \mathbf{v}_i$.

We illustrate the decoding technique with an example.

Example 7.

We construct a generator matrix for $R(1,3)$, using

$$B_3 = \begin{bmatrix} 0&1&0&1&0&1&0&1 \\ 0&0&1&1&0&0&1&1 \\ 0&0&0&0&1&1&1&1 \end{bmatrix} = \begin{bmatrix} \mathbf{v}_1 \\ \mathbf{v}_2 \\ \mathbf{v}_3 \end{bmatrix}.$$

Notice that we have ordered the columns of B_3 so that they are P_3. This B_3 differs from the B_3 in example 1 (although the codes from the corresponding generators G are equivalent). The generator matrix corresponding to B_3 here is

$$G = \begin{bmatrix} 1&1&1&1&1&1&1&1 \\ 0&1&0&1&0&1&0&1 \\ 0&0&1&1&0&0&1&1 \\ 0&0&0&0&1&1&1&1 \end{bmatrix}.$$

Using the proper ordering P_3 we construct the Hadamard matrix $H = H(2^3)$.

$$H = \begin{bmatrix} 1&1&1&1&1&1&1&1 \\ 1&-1&1&-1&1&-1&1&-1 \\ 1&1&-1&-1&1&1&-1&-1 \\ 1&-1&-1&1&1&-1&-1&1 \\ 1&1&1&1&-1&-1&-1&-1 \\ 1&-1&1&-1&-1&1&-1&1 \\ 1&1&-1&-1&-1&-1&1&1 \\ 1&-1&-1&1&-1&1&1&-1 \end{bmatrix}.$$

Suppose $\mathbf{r} = (0111\ 0110)$ is received by the decoder. Then

$$\mathbf{R} = (1, -1, -1, -1,\ 1, -1, -1, 1)$$

and by vector-matrix multiplication,

$$\hat{\mathbf{R}} = \mathbf{R}H = (-2, 2, 2, 6,\ -2, 2, 2, -2).$$

The largest component magnitude in $\hat{\mathbf{R}}$ occurs in position 4. This component corresponds to column 4 of B_3, so we set $\mathbf{u} = (110)^T$. Since this entry $\hat{\mathbf{R}}(\,(110)\,) = 6$ is positive we decode $\mathbf{r}$ to

$$\mathbf{c} = \sum_{i=1}^{3} u_i \mathbf{v}_i = 1 \cdot \mathbf{v}_1 + 1 \cdot \mathbf{v}_2 + 0 \cdot \mathbf{v}_3 = (0110\ 0110).$$

Observe that $\hat{\mathbf{R}}$ and (C) allow us to compute the distance from $\mathbf{r}$ to each of the codewords. For example, component 1 of $\hat{\mathbf{R}}$ is -2. This component corresponds to $(000)^T$. By (C), $\mathbf{r}$ is distance 5 from the vector $\mathbf{0}$ and distance 3 from the vector $\mathbf{1}$.

Working through the preceding example, we see that computing $\hat{\mathbf{R}}$ by ordinary vector-matrix multiplication, even in a small case, is time consuming. The situation can be improved considerably by using the *fast Hadamard transform*. We first require a result on the Kronecker product.

Lemma 4.3. For $A=[a_{ij}]$, $B=[b_{ij}]$, $C=[c_{ij}]$ and $D=[d_{ij}]$, the matrix product of $(A\times B)$ and $(C\times D)$ is

$$(A\times B)(C\times D)=AC\times BD.$$

Proof.

$A\times B=[a_{ij}B]$ and $C\times D=[c_{ij}D]$. Hence $(A\times B)(C\times D)=[e_{ij}]$ where

$$e_{ij}=\sum(a_{ik}B)(c_{kj}D)=\sum a_{ik}c_{kj}BD$$

using matrix block multiplication. Since $AC=[f_{ij}]$ where $f_{ij}=\sum_{k=1}^{n}a_{ik}c_{kj}$, we have

$$AC\times BD=[f_{ij}BD]=(A\times B)(C\times D).$$

□

Now using $H_2=H(2^1)$, we have the following result.

Theorem 4.4 (*Fast Hadamard Transform Theorem*). For r a positive integer,

$$H(2^r)=M_{2^r}^{(1)}\,M_{2^r}^{(2)}\cdots M_{2^r}^{(r)}$$

where $M_{2^r}^{(i)}=I_{2^{r-i}}\times H_2\times I_{2^{i-1}}$.

Proof.

The proof is by induction on r. First consider the case $r=1$. The claim is $H(2^1)=M_{2^1}^{(1)}$, and we check to see

$$M_{2^1}^{(1)} = I_{2^{1-1}}\times H_2\times I_{2^{1-1}} = H_2 = H(2^1).$$

Therefore, $H(2^1)=M_2^{(1)}$ and the result is true. Assume the result is true for $r=k$, i.e.

$$H(2^k) = M_{2^k}^{(1)}\,M_{2^k}^{(2)}\cdots M_{2^k}^{(k)}.$$

Consider $r=k+1$. Note that

$$M_{2^{k+1}}^{(i)} = I_{2^{k+1-i}}\times H_2\times I_{2^{i-1}} = I_2\times I_{2^{k-i}}\times H_2\times I_{2^{i-1}} = I_2\times M_{2^k}^{(i)},$$

using the fact that the Kronecker product is associative. Now noting that by definition,

$M_{2^{k+1}}^{(k+1)} = H_2 \times I_{2^k}$, consider

$$M_{2^{k+1}}^{(1)} M_{2^{k+1}}^{(2)} \cdots M_{2^{k+1}}^{(k+1)} = (I_2 \times M_{2^k}^{(1)})(I_2 \times M_{2^k}^{(2)}) \cdots (I_2 \times M_{2^k}^{(k)})(M_{2^{k+1}}^{(k+1)}).$$

Using Lemma 4.3 repeatedly, this is

$$= (I_2 \times M_{2^k}^{(1)} M_{2^k}^{(2)} \cdots M_{2^k}^{(k)})(H_2 \times I_{2^k})$$

$$= I_2 H_2 \times H(2^k) I_{2^k}$$

$$= H_2 \times H(2^k)$$

$$= H(2^{k+1}).$$

This completes the proof. □

Example 8.

Consider $H = H(2^3)$ and the following matrices. We use the convention that matrix entries not shown are 0.

$$M_8^{(1)} = I_4 \times H_2 \times I_1 = \begin{bmatrix} 1 & 1 & & & & & & \\ 1 & -1 & & & & & & \\ & & 1 & 1 & & & & \\ & & 1 & -1 & & & & \\ & & & & 1 & 1 & & \\ & & & & 1 & -1 & & \\ & & & & & & 1 & 1 \\ & & & & & & 1 & -1 \end{bmatrix}$$

$$M_8^{(2)} = I_2 \times H_2 \times I_2 = \begin{bmatrix} 1 & 0 & 1 & 0 & & & & \\ 0 & 1 & 0 & 1 & & & & \\ 1 & 0 & -1 & 0 & & & & \\ 0 & 1 & 0 & -1 & & & & \\ & & & & 1 & 0 & 1 & 0 \\ & & & & 0 & 1 & 0 & 1 \\ & & & & 1 & 0 & -1 & 0 \\ & & & & 0 & 1 & 0 & -1 \end{bmatrix}$$

$$M_8^{(3)} = H_2 \times I_4 = \begin{bmatrix} 1 & & & & 1 & & & \\ & 1 & & & & 1 & & \\ & & 1 & & & & 1 & \\ & & & 1 & & & & 1 \\ 1 & & & & -1 & & & \\ & 1 & & & & -1 & & \\ & & 1 & & & & -1 & \\ & & & 1 & & & & -1 \end{bmatrix}.$$

It is now easy to check that $H = H(2^3) = M_8^{(1)} M_8^{(2)} M_8^{(3)}$. Notice that each row and column of $M_{2^m}^{(i)}$ has precisely 2 non-zero terms.

Returning to example 7, we had $\mathbf{R} = (1, -1, -1, -1, 1, -1, -1, 1)$. We can now compute $\hat{\mathbf{R}} = \mathbf{R}H$ via $\mathbf{R}M_8^{(1)} M_8^{(2)} M_8^{(3)}$.

$$\mathbf{R}M_8^{(1)} = (0, 2, -2, 0,\ 0, 2, 0, -2)$$

$$(\mathbf{R}M_8^{(1)})(M_8^{(2)}) = (-2, 2, 2, 2,\ 0, 0, 0, 4)$$

$$(\mathbf{R}M_8^{(1)} M_8^{(2)})M_8^{(3)} = (-2, 2, 2, 6,\ -2, 2, 2, -2).$$

Note the order of multiplication. We choose to repeatedly multiply the vector with a matrix, and avoid multiplying two matrices directly (which would be more costly).

Let us compare the number of operations required to compute $\mathbf{R}H$ directly and as $\mathbf{R}M_{2^r}^{(1)} M_{2^r}^{(2)} \cdots M_{2^r}^{(r)}$. Let an operation be a multiplication or an addition. Suppose we multiply out $\mathbf{R}H$ by the ordinary method. Consider $\mathbf{R}$ multiplied by the first column of H. There are 2^r multiplications and $2^r - 1$ additions which are required. This is true for each column and so the number of operations to compute $\mathbf{R}H$ is

$$2^r(2^r + 2^r - 1) = 2^r(2^{r+1} - 1).$$

Now, we count the number of operations if we use the fast Hadamard transform. First we compute $\mathbf{R}M_{2^r}^{(1)}$ which will require $3(2^r)$ operations (i.e. 2 multiplications and 1 addition per column, since there are only 2 non-zero entries). We must perform this many operations for each $M_{2^r}^{(i)}$, $1 \le i \le r$. Therefore, the total number of operations is $3(2^r)r$. Hence computing $\hat{\mathbf{R}}$ via the fast Hadamard transform requires

$$\frac{3r2^r}{2^r(2^{r+1}-1)} = \frac{3r}{(2^{r+1}-1)}$$

as many operations as by the ordinary vector-matrix multiplication $\mathbf{R}H$.

$R(1,5)$ was the code used by the *Mariner* 9 spaceprobe. The code consists of 64 codewords of block length 32. Each codeword was associated with shade of grey numbered 1 through 64. A photograph taken by the craft was blocked off into a 600×600 grid. For each of the 360,000 cells the shade of grey would be determined and assigned a 32-bit codeword. This codeword was then transmitted to earth.

Since the code has distance 16, it can correct $\lfloor (16-1)/2 \rfloor = 7$ errors and detect 8 errors. $R(1,5)$ has good error-correcting capability, but has the disadvantage of relatively few codewords and a very low rate. In order to transmit colour pictures many more codewords would be necessary.

Note that to decode $R(1,5)$ by the *standard array* method would require storing information for $2^{32}/2^6 = 2^{26}$ coset leaders.

4.3 Self-Dual Codes and the Binary Golay Code†

In this section we consider a very interesting class of codes, the *self-dual* codes. From both a practical and a theoretical viewpoint, these codes have found many uses. We begin with a definition.

Definition. A linear code C is *self-orthogonal* if $C \subseteq C^\perp$.

A self-orthogonal code is one in which every code vector is orthogonal to every code vector. In such a code C, it is possible that C is smaller than $C^\perp$. If they have the same size, then it must be that $C = C^\perp$. We can easily determine if a code is self-orthogonal by studying its generator matrix G, as the following result indicates.

Lemma 4.5. A linear code C with generator matrix G is self-orthogonal if and only if $GG^T = 0$.

† This section may be omitted without loss of continuity.

Proof.

Assume $C \subseteq C^\perp$. Let $\mathbf{r}_i$ be a row of G. Then $\mathbf{r}_i \in C$, and $C \subseteq C^\perp$ implies $\mathbf{r}_i \in C^\perp$. Since G is a parity-check matrix for $C^\perp$, $G\mathbf{r}_i^T = 0$. Since this is true for every row $\mathbf{r}_i$ of G, $GG^T = 0$.

Now assume $GG^T = 0$. Let G have k rows $\mathbf{r}_i$, $1 \le i \le k$. Since the rows of G generate (form a basis for) C, $\mathbf{r}_i \in C$. Since $GG^T = 0$, $\mathbf{r}_i \cdot \mathbf{r}_j = 0$, $1 \le i,j \le k$. Suppose $\mathbf{c}_1, \mathbf{c}_2 \in C$ (not necessarily distinct). Then $\mathbf{c}_1, \mathbf{c}_2$ are linear combinations of the $\mathbf{r}_i$. Consider $\mathbf{c}_1 \cdot \mathbf{c}_2$. By the simple properties of the inner product (§3.2) and the fact that $\mathbf{r}_i \cdot \mathbf{r}_j = 0$, it follows that $\mathbf{c}_1 \cdot \mathbf{c}_2 = 0$. Hence any two codewords are orthogonal, and $C \subseteq C^\perp$. □

We now define a more specific class of codes.

Definition. A linear code C is *self-dual* if $C = C^\perp$.

Clearly, a self-dual code is self-orthogonal. However, the converse is not necessarily true.

Example 9.

The matrix

$$G = \begin{bmatrix} 1 & 0 & 0 & 1 & 0 \\ 1 & 0 & 1 & 1 & 1 \end{bmatrix}$$

is a generator for a self-orthogonal code C, because $GG^T = 0$. G does not generate a self-dual code, since C is a subspace of dimension 2, and $C^\perp$ is a subspace of dimension 3, so $C \ne C^\perp$. No (5,2)-code C can be self-dual, since for any such code C, $C^\perp$ is a (5,3)-code.

Example 10.

$$G = \begin{bmatrix} 1 & 1 & 1 & 1 & 1 & 1 & 1 & 1 \\ 1 & 0 & 0 & 1 & 1 & 0 & 1 & 0 \\ 0 & 1 & 0 & 0 & 1 & 1 & 1 & 0 \\ 0 & 0 & 1 & 1 & 0 & 1 & 1 & 0 \end{bmatrix}$$

is a generator matrix for a binary self-dual (8,4)-code C. To show that G is self-dual, we verify that $GG^T = 0$ and C has $n = 2k$. (Check that G has $k = 4$ linearly independent rows.)

Example 11.

$G = [I_6\, B]$ is a generator matrix for a self-dual (12,6)-code C over $GF(3)$, where

$$B = \begin{bmatrix} 2 & 2 & 1 & 2 & 0 & 1 \\ 0 & 2 & 2 & 1 & 2 & 1 \\ 2 & 2 & 0 & 1 & 1 & 2 \\ 1 & 0 & 1 & 1 & 1 & 1 \\ 1 & 2 & 2 & 2 & 1 & 0 \\ 1 & 2 & 1 & 0 & 2 & 2 \end{bmatrix}.$$

This follows since $GG^T = 0$ and $C^{\perp}$ has the same dimension as C. Note that to check that $GG^T = 0$ we need only check that $BB^T = -I_6$.

Example 11 illustrates the following general result for self-dual codes.

Lemma 4.6. If $G = [I_k\, B]$ is a generator matrix for a self-dual (n,k)-code over a field F, then $BB^T = -I_k$.

Proof.

Since G is self-dual, $GG^T = 0$ and

$$0 = [I_k\, B][I_k\, B]^T = I_k + BB^T$$

implying that $BB^T = -I_k$. □

Note that for binary codes with generator $G = [I_k\, B]$, we will have $BB^T = I_k$.

An important self-dual code is the binary (24,12,8) *extended Golay code*. It is the code with generator matrix $G = [I_{12}\, B]$ where

$$B = \begin{bmatrix} \cdot & 1 & 1 & 1 & 1 & 1 & 1 & 1 & 1 & 1 & 1 & 1 \\ 1 & 1 & 1 & \cdot & 1 & 1 & 1 & \cdot & \cdot & \cdot & 1 & \cdot \\ 1 & 1 & \cdot & 1 & 1 & 1 & \cdot & \cdot & \cdot & 1 & \cdot & 1 \\ 1 & \cdot & 1 & 1 & 1 & \cdot & \cdot & \cdot & 1 & \cdot & 1 & 1 \\ 1 & 1 & 1 & 1 & \cdot & \cdot & \cdot & 1 & \cdot & 1 & 1 & \cdot \\ 1 & 1 & 1 & \cdot & \cdot & \cdot & 1 & \cdot & 1 & 1 & \cdot & 1 \\ 1 & 1 & \cdot & \cdot & \cdot & 1 & \cdot & 1 & 1 & \cdot & 1 & 1 \\ 1 & \cdot & \cdot & \cdot & 1 & \cdot & 1 & 1 & \cdot & 1 & 1 & 1 \\ 1 & \cdot & \cdot & 1 & \cdot & 1 & 1 & \cdot & 1 & 1 & 1 & \cdot \\ 1 & \cdot & 1 & \cdot & 1 & 1 & \cdot & 1 & 1 & 1 & \cdot & \cdot \\ 1 & 1 & \cdot & 1 & 1 & \cdot & 1 & 1 & 1 & \cdot & \cdot & \cdot \\ 1 & \cdot & 1 & 1 & \cdot & 1 & 1 & 1 & \cdot & \cdot & \cdot & 1 \end{bmatrix}$$

and dots in B represent 0's.

Theorem 4.7. The binary extended Golay code is a self-dual (24,12,8)-code.

Proof.

Let C be the binary code with generator matrix $G = [I_{12}\, B]$. It is clear that C is a (24,12)-code (since G has full row rank and 24 columns). It is easily checked that $GG^T = 0$, so C is a self-dual binary code. It remains to prove that C has distance 8. We prove this by establishing that the minimum non-zero weight is 8.

Using the facts that C is self-dual and that every row of G has weight divisible by 4, it is not hard to prove that every codeword has weight divisible by 4 (exercise 11). If we can prove that C has no words of weight 4, then the minimum non-zero weight is 8. We observe that the first row of G has weight 12 and the others have weight 8. Hence any single row of G has weight at least 8. Label the rows of G by $\mathbf{r}_1, \mathbf{r}_2, \ldots, \mathbf{r}_{12}$ where $w(\mathbf{r}_1) = 12$ and $w(\mathbf{r}_i) = 8$, $2 \le i \le 12$. Consider a codeword

$$\mathbf{c} = \mathbf{r}_1 + \mathbf{r}_i, \;\; i \in \{2,3,\ldots,12\}.$$

It is easily checked that $w(\mathbf{c}) = 8$. Next, consider a codeword

$$\mathbf{c} = \mathbf{r}_i + \mathbf{r}_j, \;\; i \ne j, \;\; i,j \in \{2,3,\ldots,12\}.$$

If we check all such codewords, we find the weight is 8. There are $\binom{11}{2} = 55$ possibilities to check but the work can be reduced to only checking 10 pairs. If we compute

$$\mathbf{c} = \mathbf{r}_2 + \mathbf{r}_j, \ j \in \{3,4,...,12\}$$

then this will suffice, because the last 11 positions in rows 2 through 12 of G are cyclic shifts of each other. So the sum of any two rows has weight exactly 8.

Suppose now we consider codewords of the form

$$\mathbf{c} = \mathbf{r}_i + \mathbf{r}_j + \mathbf{r}_k, \ i < j < k$$

and suppose $w(\mathbf{c}) = 4$. Let $\mathbf{x}$ be the first 12 positions of $\mathbf{c}$ and $\mathbf{y}$ the last 12, and denote $\mathbf{c}$ by $(\mathbf{x},\mathbf{y})$. Since $w(\mathbf{c}) = 4$ and since $G = [I_{12}\, B]$, we have that $w(\mathbf{x}) = 3$ and $w(\mathbf{y}) = 1$. But C is self-dual, and hence $H = [B^T\, I_{12}]$ also generates C. The vector $\mathbf{c} = (\mathbf{x},\mathbf{y})$ must be a row of H since $\mathbf{c}$ must be a linear combination of rows of H and any linear combination involving 2 or more rows of H would force $w(\mathbf{y}) \geq 2$. We know that the weight of each row of H is at least 8 (by looking at G), and so $w(\mathbf{c}) \neq 4$. Hence the sum of any 3 rows of G has weight at least 8.

Can 4 distinct rows of G sum to a vector of weight 4? Let $\mathbf{c} = \mathbf{r}_i + \mathbf{r}_j + \mathbf{r}_k + \mathbf{r}_h$ be such a vector. As above, we can write $\mathbf{c} = (\mathbf{x},\mathbf{y})$. It follows that $w(\mathbf{x}) = 4$ and $w(\mathbf{y}) = 0$. But H contains no vector of this type in its row space, so $w(\mathbf{c}) \geq 8$.

Finally, suppose a codeword $\mathbf{c}$ is the sum of 5 or more rows of G. If $\mathbf{c} = (\mathbf{x},\mathbf{y})$, then $w(\mathbf{x}) \geq 5$, which forces $w(\mathbf{c}) \geq 8$. We conclude that $w(\mathbf{c}) \geq 8$ for all $\mathbf{c} \in C$. It follows that C has distance $d = 8$. □

From this result, we see that the binary extended Golay code can correct up to 3 errors and detect 4 errors simultaneously. We now present a decoding algorithm for this code, assuming at most 3 errors occur.

A Decoding Algorithm for the binary extended Golay Code.

Let $G = [I_{12}\ B]$ be the generator matrix.

Let $\mathbf{l}_1, \mathbf{l}_2, ..., \mathbf{l}_{12}$ be the columns of B and $\mathbf{r}_1, \mathbf{r}_2, ..., \mathbf{r}_{12}$ the rows.

Let $\mathbf{0}$ denote the 12-tuple of all zeros.

Let $\mathbf{x}^{(i)}, \mathbf{y}^{(i)}$ denote binary 12-tuples in which only component i is non-zero.

Let $\mathbf{r}$ be a vector received by the decoder.

(1) Compute $\mathbf{s} = G\mathbf{r}^T$.

(2) If $w(\mathbf{s}) \le 3$, then set $\mathbf{e} = (\mathbf{s}^T, \mathbf{0})$ and go to (8).

(3) If $w(\mathbf{s}+\mathbf{l}_i) \le 2$ for some $\mathbf{l}_i$, $1 \le i \le 12$, then set $\mathbf{e} = ((\mathbf{s}+\mathbf{l}_i)^T, \mathbf{y}^{(i)})$ and go to (8).

(4) Compute $B^T\mathbf{s}$.

(5) If $w(B^T\mathbf{s}) \le 3$, then set $\mathbf{e} = (\mathbf{0}, (B^T\mathbf{s})^T)$ and go to (8).

(6) If $w(B^T\mathbf{s}+\mathbf{r}_i^T) \le 2$ for some $\mathbf{r}_i$, $1 \le i \le 12$, then set $\mathbf{e} = (\mathbf{x}^{(i)}, (B^T\mathbf{s})^T+\mathbf{r}_i)$ and go to (8).

(7) At least 4 errors have occurred. Retransmit, and restart at (1).

(8) Decode $\mathbf{r}$ to $\mathbf{r} - \mathbf{e} = \mathbf{c}$.

We now give a brief outline of a proof that this decoding scheme corrects all error patterns of weight 3 or less. Note that since the code is self-dual, the generator G is also a parity-check matrix, and the 12-tuple $\mathbf{s} = G\mathbf{r}^T$ is the syndrome of the received 24-tuple $\mathbf{r}$.

Let $\mathbf{e} = (\mathbf{x},\mathbf{y})$ be an error pattern where $\mathbf{x}$ and $\mathbf{y}$ are each binary 12-tuples. Let $\mathbf{c}$ be the transmitted codeword, and let the received vector be $\mathbf{r} = \mathbf{c} + \mathbf{e}$. For $w(\mathbf{e}) \le 3$, we have the following possibilities.

(a) $w(\mathbf{x}) \le 3$, $w(\mathbf{y}) = 0$.

(b) $w(\mathbf{x}) \le 2$, $w(\mathbf{y}) = 1$.

(c) $w(\mathbf{x}) \le 1$, $w(\mathbf{y}) = 2$.

(d) $w(\mathbf{x}) = 0$, $w(\mathbf{y}) = 3$.

Since $\mathbf{r} = \mathbf{c} + \mathbf{e}$, $G\mathbf{r}^T = G\mathbf{e}^T = \mathbf{s}$. If $w(\mathbf{y}) = 0$, then $G\mathbf{e}^T = I_{12}\mathbf{x}^T = \mathbf{x}^T = \mathbf{s}$. Hence $\mathbf{e} = (\mathbf{x},\mathbf{y}) = (\mathbf{s}^T, \mathbf{0})$. This type of error pattern is found in step 2.

If $w(\mathbf{x}) \le 2$ and $w(\mathbf{y}) = 1$, then $\mathbf{y}$ has precisely one coordinate containing a 1. Let this coordinate be the ith. Then

$$G\mathbf{r}^T = I_{12}\mathbf{x}^T + B\mathbf{y}^T = I_{12}\mathbf{x}^T + \mathbf{l}_i = \mathbf{s},$$

$\mathbf{x} = (\mathbf{s}+\mathbf{l}_i)^T$ and $w(\mathbf{s}+\mathbf{l}_i) = w(\mathbf{x}) \le 2$. All error patterns of this type are found in step 3.

If $w(\mathbf{y}) = 2$ or 3 and $w(\mathbf{x}) = 0$, then $\mathbf{s} = G\mathbf{r}^T = B\mathbf{y}^T = l_i + l_j$ or $\mathbf{l}_i + \mathbf{l}_j + \mathbf{l}_h$, and since $BB^T = B^TB = I$ we have

$$\mathbf{y}^T = B^T\mathbf{s} = B^T(\mathbf{l}_i + \mathbf{l}_j) \quad \text{or} \quad B^T(\mathbf{l}_i + \mathbf{l}_j + \mathbf{l}_h).$$

It follows that $w(B^T\mathbf{s}) \le 3$, and $B^T\mathbf{s}$ is a vector with 1's in positions i and j, or i, j and h depending on the weight. The error vector is $\mathbf{e} = (\mathbf{0}, (B^T\mathbf{s})^T)$ and all such error patterns are found in step 5.

Finally, if $w(\mathbf{y}) = 2$ and $w(\mathbf{x}) = 1$, then

$$G\mathbf{r}^T = I_{12}\mathbf{x}^T + B\mathbf{y}^T = \mathbf{s},$$

and if the non-zero component in $\mathbf{x}$ is component i,

$$B^T\mathbf{s} = B^T(I_{12}\mathbf{x}^T + B\mathbf{y}^T) = B^T\mathbf{x}^T + B^TB\mathbf{y}^T = \mathbf{r}_i^T + \mathbf{y}^T.$$

Thus $\mathbf{y}^T = B^T\mathbf{s} + \mathbf{r}_i^T$ and $\mathbf{e} = (\mathbf{x}, (B^T\mathbf{s} + \mathbf{r}_i^T)^T)$. Any error pattern of this type will be found in step 6.

Example 12.

We decode the received word $\mathbf{r} = (1000\ 1000\ 0000\ 1001\ 0001\ 1101)$ as follows. Compute $G\mathbf{r}^T = (0100\ 1000\ 0000)^T = \mathbf{s}$.

Since $w(\mathbf{s}) \le 3$ we take $\mathbf{e} = (\mathbf{s}^T, \mathbf{0})$ and decode $\mathbf{r}$ to

$$\mathbf{r} - (\mathbf{s}^T, \mathbf{0}) = (1100\ 0000\ 0000\ 1001\ 0001\ 1101).$$

Note that this is the codeword which is the sum of rows 1 and 2 of G.

Example 13.

We decode $\mathbf{r} = (1000\ 0010\ 0000\ 1000\ 1101\ 0010)$ as follows.
Compute $G\mathbf{r}^T = (1011\ 1110\ 1011)^T = \mathbf{s}$.

Since $w(\mathbf{s}) > 3$ we try to find $\mathbf{l}_i$ such that $w(\mathbf{s}+\mathbf{l}_i) \le 2$.
We see after checking columns 1 through 4 of B that

$$\mathbf{s}+\mathbf{l}_4 = (0000\ 0110\ 0000)^T.$$

This has weight 2, so $\mathbf{e} = (\mathbf{s}+\mathbf{l}_4, \mathbf{y}^{(4)})$ where $\mathbf{y}^{(4)} = (0001\ 0000\ 0000)$. We decode $\mathbf{r}$ to

$$\mathbf{r}-\mathbf{e} = (1000\ 0100\ 0000\ 1001\ 1101\ 0010).$$

This is the codeword which is the sum of rows 1 and 6 of G.

Example 14.

To decode $\mathbf{r} = (1011\ 0001\ 0000\ 1100\ 1111\ 0001)$, we proceed as follows.
We compute $G\mathbf{r}^T = (1100\ 1010\ 1011)^T = \mathbf{s}$.

Since $w(\mathbf{x}) > 3$ we see if there is a column $\mathbf{l}_i$ of B such that $w(\mathbf{s}+\mathbf{l}_i) \le 2$.

Trying this we find that no such column exists. Hence we compute

$$B^T\mathbf{s} = (0101\ 1111\ 0111)^T.$$

Since $w(B^T\mathbf{s}) > 3$, we must check rows of B with $B^T\mathbf{s}$ to see if for some $\mathbf{r}_i$, $w(B^T\mathbf{s}+\mathbf{r}_i^T) \le 2$.
We see that for $i = 1$,

$$B^T\mathbf{s}+\mathbf{r}_1^T = (0010\ 0000\ 1000)^T = \mathbf{y}^T.$$

Thus the error pattern is $\mathbf{e} = (\mathbf{x},\mathbf{y})$ where $\mathbf{x} = (1000\ 0000\ 0000)$, so

$$\mathbf{e} = (1000\ 0000\ 0000\ 0010\ 0000\ 1000)$$

and we decode to

$$\mathbf{r}-\mathbf{e} = (0011\ 0001\ 0000\ 1110\ 1111\ 1001).$$

We can check that this is indeed a codeword, the sum of rows 3, 4 and 8 of G.

As mentioned in the introduction to this chapter, the Golay (24,12,8) self-dual code was used on the *Voyager* space mission to transmit pictures of Jupiter and Saturn. The code contains $2^{12} = 4096$ codewords allowing for colour transmissions. As we have seen, while the code is only 3-error correcting, its rate (½)is much better than the first order Reed-Muller code $R(1,5)$ of §4.2. Again, we mention that the suitability of a code for a particular application depends not only on the properties of the code, but also on how these match the requirements dictated by the characteristics of the transmission medium.

The *Voyager* 2 spacecraft was sent on to Uranus, and made its closest approach in January of 1986. From Uranus, the coding scheme used was a *Reed-Solomon code* (see §7.2). *Voyager* 2 is expected to encounter Neptune between June and October of 1989 (12 years after its launch).

4.4 Exercises

Hadamard matrices.

1. Let H be a $2^r \times 2^r$ matrix formed by indexing the rows and columns of H by the proper ordering P_r and placing the entries $(-1)^{\mathbf{u}\cdot\mathbf{v}}$ in row **u**, column **v**. Prove that H is a Hadamard matrix and $H = H_2 \times H_2 \times \cdots \times H_2 = H(2^r)$.

2. Let H_n be a Hadamard matrix of order $n > 1$.

   (a) Prove that n is even.

   (b) Prove that n is divisible by 4 if $n \geq 4$.

First order Reed-Muller codes.

3. Consider the first order Reed-Muller code $R(1,4)$ with generator matrix

$$G = \begin{bmatrix} 1&1&1&1&1&1&1&1&1&1&1&1&1&1&1&1 \\ 0&1&0&1&0&1&0&1&0&1&0&1&0&1&0&1 \\ 0&0&1&1&0&0&1&1&0&0&1&1&0&0&1&1 \\ 0&0&0&0&1&1&1&1&0&0&0&0&1&1&1&1 \\ 0&0&0&0&0&0&0&0&1&1&1&1&1&1&1&1 \end{bmatrix}$$

and associated Hadamard matrix $H = \begin{bmatrix} H_8 & H_8 \\ H_8 & -H_8 \end{bmatrix}$, where

$$H_8 = \begin{bmatrix} H_4 & H_4 \\ H_4 & -H_4 \end{bmatrix}, \quad H_4 = \begin{bmatrix} H_2 & H_2 \\ H_2 & -H_2 \end{bmatrix}, \quad H_2 = \begin{bmatrix} 1 & 1 \\ 1 & -1 \end{bmatrix}.$$

   (a) Compute the Hadamard transform $\hat{\mathbf{R}}$ of $\mathbf{r} = (1110\ 1110\ 0110\ 0110)$ using $\hat{\mathbf{R}} = \mathbf{R}H$, and decode **r** to the nearest codeword.

   (b) Compute $M_{16}^{(i)}$, $1 \leq i \leq 4$.

   (c) Decode **r** using part (b) and the fast Hadamard transform to compute $\hat{\mathbf{R}}$.

   (d) Decode the following vectors, using the method of either (a) or (c).

   (i) $\mathbf{r}_1 = (0000\ 1010\ 1011\ 0000)$ (ii) $\mathbf{r}_2 = (1101\ 1011\ 1101\ 0011)$

   (iii) $\mathbf{r}_3 = (1011\ 1011\ 1000\ 1001)$ (iv) $\mathbf{r}_4 = (1111\ 1000\ 0001\ 1001)$

   (v) $\mathbf{r}_5 = (1010\ 1100\ 1100\ 0011)$

4. (a) Construct the generator matrix for the first order Reed-Muller code $R(1,3)$.

   (b) Construct the matrices $M_8^{(i)}$, $1 \le i \le 3$.

   (c) Determine the Hadamard transform of the received vector $\mathbf{r} = (1011\ 0101)$, and decode $\mathbf{r}$ to the nearest codeword.

5. Consider the (n,k,d) first order Reed-Muller code $R(1,r)$.

   (a) Express n, k, and d in terms of r.

   (b) Determine the number of errors that $R(1,3)$ can correct.

   (c) Construct a generator matrix G for $R(1,3)$, with the columns corresponding to rows 2 through k of G being in proper order.

   (d) Suppose the vector $\mathbf{r} = (1001\ 1110)$ is received, and a trustworthy friend computes its Hadamard transform to be $\hat{\mathbf{R}} = (-2\ -2\ -2\ -2\ \ 2\ 2\ 2\ -6)$. Determine the transmitted codeword, assuming at most one error has occurred.

6. Determine the rate of the first order Reed-Muller code $R(1,r)$, in terms of r. How does this compare to the rate of the Hamming code of order r?

7. Determine all values of r (if any) for which the first order Reed-Muller code $R(1,r)$ is perfect.

8. Describe a method for encoding a binary information $(r+1)$-tuple $(u_0 u_1 ... u_r)$ to a codeword of $R(1,r)$ which does not require storing the generator matrix for the Reed-Muller code.

9. (a) Determine the storage requirement (in bits) for the Hadamard decoding scheme given in §4.2.

   (b) If one can tradeoff time for space, can this storage requirement be reduced? If so, to what?

10. Let $R(1,r)$ be the first order Reed-Muller code having 2^{r+1} codewords. Suppose we use the naive decoding method described in Chapter 1, i.e. we compare the received vector to each codeword.

    (a) What is the storage requirement of this scheme?

    (b) If each comparison is an operation, how many operations (in the worst case) are required to correct a received vector?

(c) Having corrected a received vector to a codeword, how does one recover the information symbols?

(d) Compare this method to that given in §4.2 using the Hadamard Transform.

Self-orthogonal and self-dual codes.

11. Prove that if the rows of a generator matrix G for a binary linear code C have weights divisible by 4 and are pairwise orthogonal, then C is self-orthogonal and every codeword has weight divisible by 4.

12. Let C be a self-orthogonal binary linear code. Prove that the weight of every codeword in C is a multiple of 4 if and only if there exists some generator matrix G for C whose rows all have weights that are multiples of 4.

13. Prove that in a binary self-orthogonal code, either all the vectors have weight divisible by 4, or half have even weight not divisible by 4, and half have weight divisible by 4.

14. Prove that an (n,k)-code C is self-dual if and only if C is self-orthogonal and $n = 2k$.

15. Decode each of the following vectors using the decoding scheme for the binary Golay code given in §4.3.

(a) $\mathbf{r}_1 = (1110\ 1000\ 0000\ 1001\ 0001\ 1101)$

(b) $\mathbf{r}_2 = (1111\ 0000\ 0000\ 0011\ 1010\ 0111)$

(c) $\mathbf{r}_3 = (0011\ 1000\ 0000\ 0100\ 1100\ 1110)$

(d) $\mathbf{r}_4 = (0000\ 0000\ 0011\ 1111\ 1101\ 1001)$

16. Let $R(1,r)$ be the first order Reed-Muller code.

(a) Prove that $R(1,r)$ is self-orthogonal for $r > 2$.

(b) Determine all values of r (if any) for which $R(1,r)$ is self-dual.

17. Give a generator matrix for a binary (10,5)-self-dual code.

18. Prove that there is no binary self-dual (10,5,4)-code.

19. Prove that there is no binary self-dual (12,6,6)-code.

20. Modify the decoding algorithm of §4.3 for a self-dual $(n,k,8)$-code over $GF(3)$.

21. Let $G = [I_6\ B]$ be a generator matrix for a (12,6)-code C over $GF(3)$ where

$$B = \begin{bmatrix} 2&2&1&2&0&1\\ 0&2&2&1&2&1\\ 2&2&0&1&1&2\\ 1&0&1&1&1&1\\ 1&2&2&2&1&0\\ 1&2&1&0&2&2 \end{bmatrix}$$

Suppose it is known that $d = 6$ for this code. Using essentially the algorithm devised in exercise 20, decode each of the following vectors:

(a) $\mathbf{r}_1 = (1100\ 2022\ 0022)$

(b) $\mathbf{r}_2 = (0201\ 0022\ 0021)$

(c) $\mathbf{r}_3 = (1000\ 0101\ 2122)$

22. Prove that the code given in exercise 21 has distance 6. (Hint: observe that we can permute rows and columns of B to get

$$B = \begin{bmatrix} 0&1&1&1&1&1\\ 2&0&2&1&1&2\\ 2&2&0&2&1&1\\ 2&1&2&0&2&1\\ 2&1&1&2&0&2\\ 2&2&1&1&2&0 \end{bmatrix}.$$

23. Let C be the (12,6,6)-code over $GF(3)$ having generator matrix $G = [I_6\, B]$ with B as given in exercise 21.

(a) Determine the number of codewords in C.

(b) Determine the number of coset leaders in a standard array for C.

(c) Determine the number of coset leaders of weight 2 in a standard array for C.

(d) Decode each of the following vectors:
(i) $\mathbf{r}_1 = (0112\ 2022\ 1010)$ (ii) $\mathbf{r}_2 = (1000\ 0101\ 2122)$

24. Let the parity check matrix for a binary (n,k)-code C be

$$H = \begin{bmatrix} 1 & 1 & 0 & 0 & 0 & 1 & 1 & 0 \\ 0 & 1 & 1 & 1 & 0 & 1 & 0 & 0 \\ 0 & 0 & 1 & 0 & 1 & 1 & 1 & 0 \\ 0 & 0 & 1 & 1 & 1 & 0 & 0 & 1 \end{bmatrix}.$$

(a) Determine n and k, and find a generator matrix G for C.

(b) Determine whether or not C is a self-dual code.

(c) Determine the distance d of C, and discuss the error correcting and detecting capabilities of C.

(d) Determine (if possible) the error pattern associated with each of the following received vectors: (i) $\mathbf{r}_1 = (0101\ 0111)$ (ii) $\mathbf{r}_2 = (1011\ 1011)$

(e) In a standard array for C are there any coset leaders having weight 3 or more?

General

25. Show that the two definitions of the Hadamard transform $\hat{\mathbf{R}}$ given in §4.2 are equivalent.

26. Let C be a binary $(n,1/2(n-1))$ code, n odd. Suppose C is a self-orthogonal code (ie. $C \subseteq C^{\perp}$). An augmented code C' is obtained from C by adding the complement of every codeword as a codeword. Prove that $C' = C^{\perp}$.

27. Let H be an Hadamard matrix of order n. Let $\mathbf{v}_1, \mathbf{v}_2, \ldots, \mathbf{v}_n$ be its rows. Consider the set of vectors $\{\mathbf{v}_1, \mathbf{v}_2, \ldots, \mathbf{v}_n, -\mathbf{v}_1, -\mathbf{v}_2, \ldots, -\mathbf{v}_n\}$. If we change all 1's to 0's and all -1's to 1's we get $2n$ binary n-tuples. Show that these n-tuples form an $[n,2n]$-code C over $A = \{0,1\}$ having distance $n/2$.

For exercises 28 through 41 we will require the following definition and notation:

Let $\mathbf{v} = (v_1, v_2, \ldots, v_t)$ and $\mathbf{w} = (w_1, w_2, \ldots, w_t)$ be binary $t = 2^m$ tuples for some $m \geq 1$. Define

$$\mathbf{vw} = (v_1 w_1, v_2 w_2, \ldots, v_t w_t) .$$

Form an array B_1 whose columns are all binary m-tuples and label the rows of B_1 as $\mathbf{u}_1, \mathbf{u}_2, \ldots, \mathbf{u}_m$ (B_1 is an $m \times 2^m$ matrix). For each k, $1 \leq k \leq m$ form an $\binom{m}{k} \times 2^m$ matrix B_k as follows: For each subset $\{i_1, i_2, \ldots, i_k\}$ of $\{1,2,\ldots,m\}$ there is a row of B_k corresponding to the vector $\mathbf{u}_{i_1}\mathbf{u}_{i_2} \cdots \mathbf{u}_{i_k}$ where this k-fold product is defined in terms of the product above. Define B_0

to be the all 1's vector and let

$$G = \begin{bmatrix} B_0 \\ B_1 \\ \cdot \\ \cdot \\ \cdot \\ B_k \end{bmatrix}.$$

Finally, we will refer to the columns of B_k and G by the corresponding columns of B_1.

28. Construct G for each of the cases $m = 2, k = 2$; $m = 3, k = 3$; $m = 4, k = 3$.

29. For each of the cases in exercise 28 determine the parameters (n,l) for the binary linear code generated by G.

30. Let B_k' be a $\binom{m}{k} \times \binom{m}{k}$ matrix obtained from B_k by using only the columns labelled with weight k columns of B_1. (There are precisely $\binom{m}{k}$ columns of this type.).

    (a) Prove that B_k' has exactly one 1 in each row and column.

    (b) Prove that you can permute the columns of B_k to obtain the identity matrix.

31. (a) For given m and $k = m$ prove that the columns of G can be permuted to obtain a matrix with all 1's down the main diagonal and 0's below the main diagonal (Hint: Permute columns of G so that weight $i-1$ columns of B_1 precede weight i columns, $1 \le i \le m$. Then permute weight i columns so that B_i' is the identity.).

    (b) Prove that the rows of G are linearly independent.

    (c) For given m and any k, $0 \le k \le m$, prove that G is a generator matrix for a $(2^m, l)$-binary code where $l = \sum_{i=0}^{k} \binom{m}{i}$.

    (The code generated by G is called the k'th *order Reed-Muller code* of length 2^m and is denoted $R(k,m)$. This generalizes the results at the beginning of this chapter.)

Exercises 32-38 deal with the distance of $R(k,m)$ codes.

32. Determine the distance in $R(k,3)$ for $k = 1,2$ and 3.

33. Show that the weight of any row in B_k is 2^{m-k}.

(Hint. If there is a 1 in row $\mathbf{u}_{i_1}\mathbf{u}_{i_2}\cdots\mathbf{u}_{i_k}$ and column c then there must be a 1 in vector $\mathbf{u}_{i_j}$ in column c for each j, $1 \le j \le k$.)

34. Permute the columns of G, a generator matrix for $R(k,m)$, such that row $\mathbf{u}_1$ of B_1 has the form $\mathbf{u}_1 = (\mathbf{u},\mathbf{w})$ where $\mathbf{u}$ is the $\mathbf{0}$ vector of length 2^{m-1} and $\mathbf{w}$ is the vector of all 1's of length 2^{m-1}. Denote the resulting matrix by G'. Let G_1' be the matrix obtained from G' by taking the rows of G' and the first 2^{m-1} columns. Let G_2' be the matrix obtained from G' by taking the rows of G' and the last 2^{m-1} columns.

    (a) Prove that the rows of G_1' span a $R(k,m-1)$. (Notice that G_1' is not a generator matrix since it has repeated rows.)

    (b) Prove that the rows of G_2' span a $R(k-1,m-1)$.

35. Using exercises 33 and 34 and mathematical induction on m, prove that the distance of $R(k,m)$ is 2^{m-k}.

36. For $m = 2t + 1 > 0$, determine the rate of an $R(t,m)$ code.

37. From exercise 33 we see that every row in B_k, $0 \le k \le m-1$, has even weight.

    (a) Show that any row from B_i is orthogonal with any row from B_j for $0 \le i \le m-1$, and $j \le m-i-1$.

    (b) Using (a) deduce that the dual of $R(k,m)$ is precisely $R(m-1-k, m)$.

38. Compute the weight enumerator of $R(m-1,m)$.

39. Compute the weight enumerator for the binary linear code with block length 16 generated by B_2.

40. Compute the weight enumerator for the binary linear code with block length 16 generated by B_3.

41. Let $\mathbf{v}$ and $\mathbf{w}$ be two distinct rows of B_k. Determine the weight of $\mathbf{v}+\mathbf{w}$.

Chapter 5

CYCLIC CODES

5.1 Introduction

One of the most important classes of linear codes is the class of *cyclic codes*. In general these codes are much easier to implement and hence have great practical importance. From an algebraic viewpoint they are also of considerable interest. Many of the codes we have already investigated are cyclic or derived from cyclic codes. We begin our study by defining a *cyclic subspace.*

Definition. A subspace S of $V_n(F)$ is a *cyclic subspace* if whenever $(a_1a_2...a_{n-1}a_n) \in S$ then $(a_na_1a_2...a_{n-1}) \in S$.

In other words, S is a subspace and for each vector $\mathbf{a} \in S$, every cyclic shift of $\mathbf{a}$ is also in S.

Definition. A linear code C is a *cyclic code* if C is a cyclic subspace.

Since cyclic codes are linear, a cyclic code necessarily contains the codeword $\mathbf{0}$ and is closed under addition.

Example 1.

(i) $S = \{(0000)\} \subseteq V_4(Z_2)$ is trivially a cyclic subspace.

(ii) $S = \{(0000),(1111)\} \subseteq V_4(Z_2)$ is a cyclic subspace.

(iii) $S = \{(0000000), (1011100), (0101110), (0010111), (1110010), (0111001), (1001011), (1100101)\}$ is a cyclic subspace in $V_7(Z_2)$.

(iv) The following subset of $V_4(Z_2)$ is not a cyclic subspace.
$S = \{(0000), (1001), (1100), (0110), (0011), (0111), (1011), (1101)\}$.
Every cyclic shift of (0111) is not in S. In fact, S is not even a subspace, as it is not closed under vector addition.

(v) $S = \{(000), (210), (021), (102), (201), (120), (012), (222), (111)\}$ is a cyclic subspace of dimension 2 in $V_3(Z_3)$.

There are several questions which arise.

1. How can cyclic subspaces be constructed?
2. For a given value of k, can a k-dimensional cyclic subspace of $V_n(F)$ be constructed?
3. How many cyclic subspaces does $V_n(F)$ contain?
4. Which vectors in a cyclic subspace S have the property that the vector and its cyclic shifts will generate all of S?

With respect to this last question, consider for example the 4-dimensional subspace S of $V_6(Z_2)$ generated by the vectors $\mathbf{v}_1 = (111000)$, $\mathbf{v}_2 = (011100)$, $\mathbf{v}_3 = (001110)$ and $\mathbf{v}_4 = (000111)$. This basis can be obtained from $\mathbf{v}_1$ and its three cyclic shifts. The vector $\mathbf{v} = (101010)$ is also an element of S, but the cyclic shifts of $\mathbf{v}$ generate only a subspace of S of dimension 2.

These questions and many more can be best answered by introducing additional algebraic structure to $V_n(F)$. We do so in the next section.

5.2 Rings and Ideals

Let us first discuss some concepts briefly mentioned in Chapter 2.

Definition. A *commutative ring with unity* $(R,+,*)$ is an algebraic structure consisting of a set of elements R together with two binary operations denoted + and * which satisfy the following properties for all $a, b, c \in R$.

(i) $(R,+)$ is an abelian group with identity e.

(ii) $(a*b)*c = a*(b*c)$.

(iii) $(a+b)*c = a*c + b*c$ and $c*(a+b) = c*a + c*b$.

(iv) There exists an element $1 \in R$ such that $a*1 = 1*a = a$.

(v) $a*b = b*a$.

Since the only rings which we will consider will be commutative rings with unity, for our purposes we will refer to these as simply *rings*.

Example 2.

(i) The set of all integers under addition and multiplication forms a ring $(Z,+,*)$, usually denoted simply Z.

(ii) The set of integers modulo a positive integer n form a ring, usually denoted Z_n.

(iii) The set of all polynomials in an indeterminate x with coefficients from a field F form a ring. This ring is usually denoted $F[x]$. The two operations are the standard polynomial addition and multiplication.

From (iii) we may construct a ring with a finite number of elements as follows. Given any non-zero polynomial $f(x) \in F[x]$, define two polynomials $h(x)$, $g(x) \in F[x]$ to be congruent modulo $f(x)$ if (and only if) $f(x)$ divides $h(x) - g(x)$, i.e. $h(x)$ and $g(x)$ leave the same remainder when divided by $f(x)$. Then as in Chapter 2, the congruence relation partitions $F[x]$ into equivalence classes, with the equivalence class containing $g(x)$ denoted $[g(x)]$ and defined as

$$[g(x)] = \{h(x): h(x) \equiv g(x) \pmod{f(x)}\}.$$

Let $R = F[x]/(f(x))$ be the set of equivalence classes, i.e.

$$F[x]/(f(x)) = \{[g(x)]: g(x) \in F[x]\}.$$

Define addition and multiplication of equivalence classes in the natural way, by the rules

$$[g(x)] + [h(x)] = [g(x) + h(x)]$$

and

$$[g(x)]*[h(x)] = [g(x)*h(x)].$$

Then $(R,+,*)$ is a ring called the *ring of polynomials over F modulo* $f(x)$. Any polynomial in the class $[g(x)]$ can be used to represent the class, and we usually take the polynomial of least degree as the representative. Then the equivalence classes are represented by all polynomials in $F[x]$ of degree less than the degree of $f(x)$, corresponding to all possible remainders after dividing by $f(x)$. When the context is clear, we omit the square brackets for the class and write only the representative, $g(x)$.

Example 3.

As a particular example of this ring, let us consider $Z_2[x]/(f(x))$ where $f(x) = x^3 + 1$.

$$Z_2[x]/(x^3+1) = \{[0], [1], [x], [1+x], [x^2], [1+x^2], [x+x^2], [1+x+x^2]\}$$

As an example of addition,

$$[x] + [1+x+x^2] = [x+1+x+x^2] = [1+x^2].$$

As an example of multiplication in this ring,

$$[1+x^2]*[1+x+x^2] = [(1+x^2)*(1+x+x^2)] = [1+x+x^3+x^4] = [0]$$

since $x^4+x^3+x+1 \equiv 0 \pmod{f(x)}$. We observe that $x^3 + 1 \equiv 0 \pmod{f(x)}$ implies $x^3 \equiv 1 \pmod{f(x)}$, and hence wherever we see x^3 in an expression we can replace it by 1.

We now define an important substructure of a ring.

Definition. Let $(R,+,*)$ be a ring. A nonempty subset I of R is called a *ideal* of the ring if

(i) $(I,+)$ is a group, and

(ii) $i*r \in I$ for all $i \in I$ and all $r \in R$.

Ideals will play a fundamental role in our study of cyclic subspaces in $V_n(F)$. In §5.3 we will establish a 1-1 correspondence between ideals in the ring of polynomials over F modulo x^n-1 and cyclic subspaces in $V_n(F)$. This permits us to study cyclic codes through the rich theory of polynomials.

One simple way to construct an ideal is as follows. Take any non-zero $g \in R$ and form the set

$$I = \{g*r:\ r \in R\}.$$

It is easy to verify that I is an ideal. It is called the *ideal generated by* g. It is not always possible to construct all ideals of a ring in this fashion. When the ring R has the property that for any ideal I of R there exists an element $g \in I$ such that $I = \{g*r:\ r \in R\}$, then R is called a *principal ideal ring*. We now prove that $F[x]$ and $F[x]/(f(x))$ are such rings.

Theorem 5.1. $F[x]$ **is a principal ideal ring.**

Proof.

Let I be an ideal of $F[x]$. If $I = \{0\}$ then I is the ideal generated by 0. Otherwise, let $g(x)$ be a monic polynomial of least degree in I, i.e. the highest power of x in $g(x)$ has coefficient 1. We prove that $g(x)$ generates I. Consider any $h(x) \in I$. By the division algorithm for polynomials,

$$h(x) = q(x)g(x) + r(x)$$

where $r(x) = 0$ or $\deg r(x) < \deg g(x)$. Since $g(x) \in I$, it follows from property (ii) that $q(x)g(x) \in I$, and by (i), $h(x) - q(x)g(x) \in I$ so $r(x) \in I$. Since $g(x)$ is a polynomial of least degree in I, we must have $r(x) = 0$ and thus $g(x)$ divides $h(x)$. This establishes that $g(x)$ generates I and $F[x]$ is a principal ideal ring. □

Theorem 5.2. $F[x]/(f(x))$ **is a principal ideal ring.**

Proof.

Let I be an ideal in R. If $I = \{[0]\}$, then I is generated by 0. Otherwise, let $g(x)$ be a monic polynomial of least degree which represents some class in I. Then $[g(x)] \in I$. Let $[h(x)] \in I$. By the division algorithm,

$$h(x) = q(x)g(x) + r(x)$$

where $\deg r(x) < \deg g(x)$ or $r(x) = 0$. Therefore

$$[h(x)] = [q(x)g(x) + r(x)] = [q(x)g(x)] + [r(x)].$$

Since $[q(x)g(x)] \in I$, it follows that $[h(x)] - [q(x)g(x)] \in I$ and hence $[r(x)] \in I$. This implies $r(x) = 0$, by the choice of $g(x)$. Hence $g(x)$ divides $h(x)$, and $g(x)$ generates I. □

Example 4.

Consider $R=Z_2[x]/(f(x))$ where $f(x) = x^6 + 1$, and the set

$$I = \{0,\ 1+x^2+x^4,\ x+x^3+x^5,\ 1+x+x^2+x^3+x^4+x^5\}.$$

I is an ideal in R. It is easy to verify that $(I,+)$ is a group (property (i)), and somewhat more time-consuming, but no harder, to verify property (ii). I is in fact the ideal generated by $g(x) = 1 + x^2 + x^4$.

We shall see that the concept of an ideal is central to the study of cyclic subspaces.

5.3 Ideals and Cyclic Subspaces

We know that $V_n(F)$ is an abelian group under vector addition. We now endow $V_n(F)$ with the properties of a ring by defining a multiplication between any two vectors. The easiest way to do this is to associate a polynomial in $F[x]$ with each vector in $V_n(F)$. If $\mathbf{v} = (a_0 a_1 \ldots a_{n-1})$, then let

$$v(x) = a_0 + a_1 x + \cdots + a_{n-1} x^{n-1}.$$

In order to define a vector multiplication we select the polynomial $f(x) = x^n - 1 \in F[x]$. We could in fact use any polynomial of degree n in $F[x]$ to define a multiplication, but we shall see that for our purposes $f(x) = x^n - 1$ is most suitable. Now, for $\mathbf{v}_1, \mathbf{v}_2 \in V_n(F)$, let $v_1(x)v_2(x) = v(x)$ where $v(x)$ is the polynomial of least degree in the equivalence class $[v_1(x)v_2(x)]$ of $F[x]/(f(x))$. In other words, $v(x)$ is the remainder polynomial when $v_1(x)v_2(x)$ is divided by $f(x)$.

With multiplication defined in this way, and with our association between polynomials and vectors, we can essentially think of $V_n(F)$ and $F[x]/(f(x))$ interchangeably. Thus an element of $V_n(F)$ may be considered as an n-tuple over F or a polynomial of degree at most $n-1$ over F. The components of the n-tuple represent the coefficients of the polynomial, from low order to high order.

Let us explain the choice of the polynomial $f(x) = x^n - 1$ used to define the multiplication. Consider $\mathbf{v} = (a_0 a_1 \ldots a_{n-1}) \in V_n(F)$.

$$v(x) = a_0 + a_1 x + \cdots + a_{n-1} x^{n-1}$$

$$xv(x) = a_0 x + a_1 x^2 + \cdots + a_{n-1} x^n.$$

Now $x^n - 1 \equiv 0 \pmod{f(x)}$ implies $x^n \equiv 1 \pmod{f(x)}$. Thus

$$xv(x) = a_{n-1} + a_0 x + a_1 x^2 + \cdots + a_{n-2} x^{n-1} = (a_{n-1} a_0 a_1 \ldots a_{n-2}),$$

and we see that multiplication by x corresponds to a cyclic shift of $\mathbf{v}$. With this observation we can establish a fundamental theorem in our study of cyclic codes.

Theorem 5.3. A non-empty subset S of $V_n(F)$ is a cyclic subspace if and only if the set of polynomials I associated with S is an ideal in the ring R of polynomials associated with $V_n(F)$.

Proof.

Assume S is a cyclic subspace. To prove I is an ideal, we must verify the two properties of the definition. The first is easy. That $(I,+)$ is an abelian group follows from S being a subspace. To prove the second, that I is closed under multiplication by elements from R, we proceed as follows. Let

$$v = (a_0 a_1 \ldots a_{n-1}) \in S.$$

Then by our association,

$$v(x) \in I.$$

Since S is cyclic, $(a_{n-1} a_0 \ldots a_{n-2}) \in S$, and thus $xv(x) \in I$. Since S is a vector space, if $v \in S$ then $\lambda v \in S$ for all $\lambda \in F$. We conclude that $\lambda x^i v(x) \in I$ for all $\lambda \in F$, $0 \le i \le n-1$. Thus $a(x)v(x) \in I$ where $a(x) = \sum_{i=0}^{n-1} \lambda_j x^j$ is any element of the ring R, and the second property is verified. Hence I is an ideal.

To prove the converse, assume that I is an ideal of R. We must prove that S is a vector space, and that S is a cyclic subspace. That S is a vector space follows from the facts that $(I,+)$ is a group and that scalar multiplication in S corresponds to multiplication by constant polynomials. It is easily verified that S now satisfies all properties of a vector space. That S is a cyclic subspace follows from the fact that I is closed under polynomial multiplication by all $r \in R$, and in particular for $r(x) = x$, and we have seen that multiplication by x corresponds to a cyclic shift of the vector corresponding to the multiplicand. □

In Theorem 5.2 we proved that every ideal in $V_n(F)$ is generated by some element of the ideal. We now make this more explicit. Recall that $f(x) = x^n - 1$ is the polynomial we chose in order to define a vector multiplication.

Theorem 5.4. Let I be a non-zero ideal in $V_n(F)$ and let $g(x)$ be a monic polynomial of least degree which represents some class of I. Then $[g(x)]$ (or less formally $g(x)$) generates I and $g(x)$ divides $f(x) = x^n - 1$.

Proof.

That $[g(x)]$ generates I follows from the proof of Theorem 5.2. We must prove $g(x)$ divides $f(x)$. Let

$$f(x) = h(x)g(x) + r(x)$$

where $\deg r(x) < \deg g(x)$ or $r(x) = 0$. Then

$$[f(x)] = [h(x)g(x) + r(x)] = [h(x)][g(x)] + [r(x)].$$

Since $[f(x)] = [0]$ we have

$$[r(x)] = [-h(x)][g(x)] \in I.$$

From the degree constraint on $r(x)$, and the choice of $g(x)$, it follows that $r(x) = 0$ and $g(x)$ divides $f(x)$. □

We now prove that this $g(x)$ is unique.

Theorem 5.5. There is a unique monic polynomial of least degree which generates a non-zero ideal I of $V_n(F)$.

Proof.

Suppose both $g(x)$ and $h(x)$ are monic polynomials of least degree in an ideal I. That is, $\deg g(x) = \deg h(x)$ and $\deg g(x)$ is a minimum in I. Then $g(x)$ generates I and so does $h(x)$. Since $g(x)$ generates I and $h(x) \in I$, then $h(x) = a(x)g(x)$ for some polynomial $a(x)$. Since $h(x)$ and $g(x)$ have the same degree and are monic, $a(x) = 1$. Therefore $h(x) = g(x)$, and the proof is complete. □

Having proven this, we will refer to the unique monic polynomial of least degree in an ideal I as the *generator polynomial.* We note that an ideal may contain many elements which will generate the ideal, but there is only one such monic polynomial of least degree, which we single out as the generator polynomial.

Theorem 5.4 proves that the generator polynomial of an ideal I must divide $f(x) = x^n - 1$. Consider the converse of this statement. Suppose $h(x)$ is a monic divisor of $f(x)$ and we form the ideal $I = \{a(x)h(x): a(x) \in R\}$ where $R = F[x]/(x^n - 1)$, and all polynomials in I are reduced mod $x^n - 1$. Is $h(x)$ the generator polynomial of I? Certainly $h(x)$ is a generator for the ideal, but is it the monic polynomial of least degree in I? We now see that this is indeed the case.

Theorem 5.6. Let $h(x)$ be a monic divisor of $f(x) = x^n - 1$. Then $h(x)$ is the generator polynomial of the ideal $I = \{a(x)h(x): a(x) \in R\}$ of $R = F[x]/(x^n - 1)$.

Proof.

Let $g(x)$ be a monic polynomial of least degree in I. Then $g(x)$ is the generator polynomial of I and by Theorem 5.4 $g(x)$ divides $f(x)$. Since $g(x) \in I$, there exists a polynomial $a(x)$ such that

$$[g(x)] = [a(x)h(x)]$$

and

$$g(x) = a(x)h(x) + l(x)f(x)$$

for some polynomial $l(x)$. Since $h(x)$ divides $f(x)$, $h(x)$ divides $g(x)$. Also, $g(x)$ divides $h(x)$ since $g(x)$ is the generator. Both are monic. We conclude that $g(x) = h(x)$. □

Combining the previous results of this section, we obtain the following important consequence.

Theorem 5.7. There is a $1-1$ correspondence between cyclic subspaces of $V_n(F)$ and monic polynomials $g(x) \in F[x]$ which divide $f(x) = x^n - 1$.

Example 5.

Consider $V_7(Z_2)$ and $f(x) = x^7 - 1$. A complete factorization of $f(x)$ over Z_2 is

$$x^7 - 1 = (x+1)(x^3+x^2+1)(x^3+x+1).$$

The monic divisors of $f(x)$ are

$$g_1(x) = 1$$

$$g_2(x) = x + 1$$

$$g_3(x) = x^3 + x^2 + 1$$

$$g_4(x) = x^3 + x + 1$$

$$g_5(x) = (x+1)(x^3 + x^2 + 1)$$

$$g_6(x) = (x+1)(x^3+x+1)$$

$$g_7(x) = (x^3+x^2+1)(x^3+x+1)$$

$$g_8(x) = f(x).$$

$g_6(x)$ generates the cyclic subspace

$$S = \{(0000000), (1011100), (0101110), (0010111), (1001011), (1100101), (1110010), (0111001)\}.$$

$g_7(x)$ generates the cyclic subspace

$$S = \{(0000000), (1111111)\}.$$

$V_7(Z_2)$ contains precisely 8 cyclic subspaces.

The theory we have developed so far answers questions 1, 3 and 4 given at the beginning of this chapter. We summarize with the following statement.

Theorem 5.8. Let $f(x) = x^n - 1$ have factorization $f(x) = p_1^{a_1}(x)p_2^{a_2}(x) \cdots p_t^{a_t}(x)$ where $p_i(x)$ are distinct irreducible monic polynomials over F, and a_i are positive integers, $1 \le i \le t$. Then $V_n(F)$ contains

$$(a_1+1)(a_2+1) \cdots (a_t+1)$$

cyclic subspaces.

Since the factorization of $x^n - 1$ into irreducible factors is so important to the study of cyclic subspaces, we include Table 3 giving the factorization of $x^n - 1$ over $GF(2)$ for small values of n. The factorization of $x^n - 1$ over $GF(q)$ is discussed in §5.8.

We have seen that if $g(x)$ is a monic divisor of $f(x) = x^n - 1$ over F then $g(x)$ is the generator polynomial of a cyclic subspace S. We now show that the degree of $g(x)$ determines the the dimension of the subspace it generates.

Table 3. Factorization of x^n-1 over $GF(2)$

n	factorization
1	$1+x$
2	$(1+x)^2$
3	$(1+x)(1+x+x^2)$
4	$(1+x)^4$
5	$(1+x)(1+x+x^2+x^3+x^4)$
6	$(1+x)^2(1+x+x^2)^2$
7	$(1+x)(1+x+x^3)(1+x^2+x^3)$
8	$(1+x)^8$
9	$(1+x)(1+x+x^2)(1+x^3+x^6)$
10	$(1+x)^2(1+x+x^2+x^3+x^4)^2$
11	$(1+x)(1+x+\cdots+x^{10})$
12	$(1+x)^4(1+x+x^2)^4$
13	$(1+x)(1+x+\cdots+x^{12})$
14	$(1+x)^2(1+x+x^3)^2(1+x^2+x^3)^2$
15	$(1+x)(1+x+x^2)(1+x+x^2+x^3+x^4)(1+x+x^4)(1+x^3+x^4)$
16	$(1+x)^{16}$
17	$(1+x)(1+x+x^2+x^4+x^6+x^7+x^8)(1+x^3+x^4+x^5+x^8)$
18	$(1+x)^2(1+x+x^2)^2(1+x^3+x^6)^2$
19	$(1+x)(1+x+x^2+\cdots+x^{18})$
20	$(1+x)^4(1+x+x^2+x^3+x^4)^4$
21	$(1+x)(1+x+x^2)(1+x^2+x^3)(1+x+x^3)(1+x^2+x^4+x^5+x^6)(1+x+x^2+x^4+x^6)$
22	$(1+x)^2(1+x+x^2+\cdots+x^{10})^2$
23	$(1+x)(1+x+x^5+x^6+x^7+x^9+x^{11})(1+x^2+x^4+x^5+x^6+x^{10}+x^{11})$
24	$(1+x)^8(1+x+x^2)^8$
25	$(1+x)(1+x+x^2+x^3+x^4)(1+x^5+x^{10}+x^{15}+x^{20})$.

Theorem 5.9. Let $g(x)$ be a monic divisor of x^n-1 over F having degree $n-k$. Then $g(x)$ is the generator polynomial for a cyclic subspace of $V_n(F)$ of dimension k.

Proof.

Let S be the cyclic subspace of $V_n(F)$ generated by $g(x)$, and consider the set of vectors corresponding to

$$B = \{g(x), x^1 g(x), \ldots, x^{k-1} g(x)\}.$$

We claim that B is a basis for S. To prove this, we show that the set B is linearly independent and spans all of S.

Consider any linear combination such that

$$\sum_{i=0}^{k-1} \lambda_i x^i g(x) = 0, \quad \lambda_i \in F.$$

The only term which contains x^{n-1} is $\lambda_{k-1} x^{k-1} g(x)$. Since the degree of $x^{k-1}g(x)$ is less than n, $\lambda_{k-1} = 0$. By similar reasoning we get $\lambda_i = 0$, $0 \le i \le k-1$, and thus B is a linearly independent set.

Now consider any element $h(x) \in S$. Since $g(x)$ is the generator of S

$$h(x) = a(x)g(x)$$

for some polynomial $a(x)$. Without loss of generality, we may assume $a(x)$ to be of degree less than k, since otherwise, $\deg a(x)g(x) \ge n$ and for some $q(x)$,

$$a(x)g(x) = q(x)(x^n - 1) + r(x)$$

where $\deg r(x) \le n-1$ or $r(x)=0$. Now $g(x)$ divides $x^n - 1$, so $g(x)$ divides $r(x)$ and $r(x) = a'(x)g(x)$ for some $a'(x)$, $\deg a'(x) \le k-1$. We can then use $a'(x)$ in place of $a(x)$. Therefore without loss of generality $a(x)$ can be written

$$a(x) = \sum_{i=0}^{k-1} \lambda_i x^i.$$

Hence

$$h(x) = \sum_{i=0}^{k-1} \lambda_i x^i g(x)$$

and B spans the space S. Since B is a linearly independent set that spans S, B is a basis for S and S has dimension k. □

This provides us with the answer to question 2 which was asked in §5.1. To construct a cyclic subspace of dimension k in $V_n(F)$, i.e. a cyclic (n,k)-code over F, we require a monic divisor of $x^n - 1$ which has degree $n-k$.

Example 6.

Suppose we wish to construct a binary cyclic (7,4)-code. Since $g(x) = x^3 + x + 1$ is a monic divisor of $x^7 - 1$, $g(x)$ generates a cyclic subspace in $V_7(Z_2)$ of dimension 4.

Example 7.

Suppose we wish to construct a binary cyclic (15,9)-code. Since $g(x) = (1+x+x^2)(1+x+x^4)$ is a monic divisor of $x^{15} - 1$ over Z_2 and $g(x)$ has degree 6, $g(x)$ generates a cyclic subspace of dimension 9.

5.4 Generator Matrices and Parity-Check Matrices

Let $g(x)$ be the generator of a cyclic (n,k)-code C over F. From the previous section we know that $g(x)$ will have degree $n-k$, and each codeword in C will have the form $a(x)g(x)$. From the proof of Theorem 5.9, we may assume $\deg a(x) \le k-1$. Knowing this, we let the message space consist of all polynomials over F of degree less than or equal to $k-1$. The channel encoder then encodes the message polynomial $a(x)$ to the codeword $a(x)g(x)$. Since $a(x)g(x)$ has degree at most $n-1$, no reduction modulo $f(x) = x^n - 1$ is necessary. A generator matrix for C is then given in terms of $g(x)$ by

$$G = \begin{bmatrix} g(x) \\ x^1 g(x) \\ x^2 g(x) \\ \cdot \\ \cdot \\ x^{k-1} g(x) \end{bmatrix}.$$

Example 8.

Consider $f(x) = x^7 - 1$. $g(x) = 1 + x + x^3$ generates a (7,4)-cyclic code over Z_2. The message space consists of all polynomials over Z_2 of degree at most 3. The polynomial $a(x) = 1 + x^2 + x^3$ encodes to the codeword polynomial $a(x)g(x) = 1 + x + x^2 + x^3 + x^4 + x^5 + x^6$. In terms of binary vectors, the message 4-tuple (1011) encodes to the codeword (1111111). A generator matrix is

$$G = \begin{bmatrix} g(x) \\ x^1 g(x) \\ x^2 g(x) \\ x^3 g(x) \end{bmatrix} = \begin{bmatrix} 1&1&0&1&0&0&0 \\ 0&1&1&0&1&0&0 \\ 0&0&1&1&0&1&0 \\ 0&0&0&1&1&0&1 \end{bmatrix}.$$

To encode $\mathbf{a} = (1011)$, compute $\mathbf{a}G = (1111111)$.

Now that we have a polynomial approach to describe a cyclic code C, we consider the related polynomial representation of the dual code $C^{\perp}$ of C. We shall see that $C^{\perp}$ is a cyclic code if C is cyclic.

Consider $h(x) = f(x)/g(x)$ where $f(x) = x^n - 1$ and $g(x)$ is a monic divisor of $f(x)$, and $g(x)$ generates C. If $\deg g(x) = n-k$, then $\deg h(x) = k$. Since $g(x)$ is monic, so is $h(x)$. It follows that $h(x)$ generates a cyclic code C' of dimension $n-k$. Now consider $c_1(x) = a_1(x)g(x) \in C$ and $c_2(x) = a_2(x)h(x) \in C'$. Then

$$c_1(x)c_2(x) = a_1(x)g(x)a_2(x)h(x) = a_1(x)a_2(x)f(x) \equiv 0 \ (\bmod f(x)).$$

Therefore, using polynomial multiplication mod $f(x)$, any codeword from C multiplies with any codeword from C' to give the polynomial 0, corresponding to the codeword $\mathbf{0}$. Does this imply that C' is the dual code of C? The answer is no, because the product of two polynomials in $F[x]/(f(x))$ being 0 does not imply that the corresponding vectors in $V_n(F)$ are orthogonal, i.e. have inner product 0 over F. However, we now show that the codes C' and $C^{\perp}$ are closely related – they are equivalent codes.

Consider two vectors

$$\mathbf{a} = (a_0 a_1 ... a_{n-1}) \in C$$

and

$$\mathbf{b} = (b_0 b_1 ... b_{n-1}) \in C',$$

and the product

$$a(x)b(x) = \left(\sum_{i=0}^{n-1} a_i x^i\right)\left(\sum_{i=0}^{n-1} b_i x^i\right) \equiv \sum_{i=0}^{n-1} c_i x^i \pmod{x^n - 1}$$

for some $c_i \in F$. The constant term in this product is

$$c_0 = a_0b_0 + a_1b_{n-1} + a_2b_{n-2} + \cdots + a_{n-1}b_1,$$

since $x^n \equiv 1 \pmod{f(x)}$. Now c_0 can be written as the inner product

$$c_0 = \mathbf{a}\cdot\overline{\mathbf{b}},$$

where $\overline{\mathbf{b}}$ is the n-tuple obtained from $\mathbf{b}$ by cyclically shifting $\mathbf{b}$ one position to the left and then reversing the order of components. Observe that multiplying $\sum_{i=0}^{n-1} c_i x^i$ by x^{n-t} results in c_t being the constant term, and hence the constant term in the product

$$a(x)\,(x^{n-t}\,b(x)).$$

Therefore

$$c_t = \mathbf{a}\cdot\overline{\mathbf{b}},$$

where $\overline{\mathbf{b}}$ is now the n-tuple associated with the polynomial $x^{n-t}b(x)$. In terms of $b(x)$, $\overline{\mathbf{b}}$ is the vector obtained by cyclically shifting $\mathbf{b}$ $t+1$ positions to the left and then reversing the order of the components.

Example 9.

Consider $n = 3$, and the vectors $\mathbf{a} = (a_0a_1a_2)$, $\mathbf{b} = (b_0b_1b_2)$. Then modulo x^3-1,

$$a(x)b(x) = (a_0+a_1x+a_2x^2)(b_0+b_1x+b_2x^2)$$

$$= (a_0b_0+a_1b_2+a_2b_1) + (a_0b_1+a_1b_0+a_2b_2)x + (a_0b_2+a_1b_1+a_2b_0)x^2.$$

The coefficient of x^0 is $\mathbf{a}\cdot(b_0b_2b_1)$. $(b_0b_2b_1)$ is obtained from $\mathbf{b}$ by shifting $\mathbf{b}$ cyclically $3-0-1 = 2$ positions to $(b_1b_2b_0)$ and reordering the coefficients last to first, yielding $(b_0b_2b_1)$.

Now since $a(x)b(x) \equiv 0 \pmod{x^n - 1}$, the coefficient of each power of x must be 0. From the discussion above, this implies that $\mathbf{a}\cdot\mathbf{c} = 0$ (i.e. $\mathbf{a}$ and $\mathbf{c}$ are orthogonal) where $\mathbf{c}$ is any cyclic shift of the vector obtained from $\mathbf{b}$ by reversing the components of $\mathbf{b}$. Since $h(x)$ generates C', $\{h(x), xh(x),\ldots, x^{n-k-1}h(x)\}$ is a basis for C' and

$$G' = \begin{bmatrix} h(x) \\ x^1 h(x) \\ \cdot \\ \cdot \\ x^{n-k-1} h(x) \end{bmatrix}$$

is a generator matrix for C'. Now G' generates the code C', which has the same dimension $n-k$ as $C^\perp$. Furthermore, taking $b(x)$ in the above arguments to be $h(x)$ itself, we see that the reverse of every vector in C' is orthogonal to every vector in C. It follows that if we reorder the columns of G' last to first, we obtain a matrix H which generates $C^\perp$, and hence is a parity-check matrix for C.

Example 10.

Suppose we wish to construct a generator matrix and a parity-check matrix for a (7,4)-binary cyclic code. Since $g(x) = 1 + x + x^3$ divides $f(x) = x^7 - 1$, $g(x)$ generates a (7,4)-cyclic code C. $h(x) = f(x)/g(x) = 1 + x + x^2 + x^4$ generates a (7,3)-cyclic code C'. A generator matrix for C is

$$G = \begin{bmatrix} g(x) \\ x^1 g(x) \\ x^2 g(x) \\ x^3 g(x) \end{bmatrix} = \begin{bmatrix} 1&1&0&1&0&0&0 \\ 0&1&1&0&1&0&0 \\ 0&0&1&1&0&1&0 \\ 0&0&0&1&1&0&1 \end{bmatrix}.$$

A generator matrix for C' is

$$G' = \begin{bmatrix} h(x) \\ x^1 h(x) \\ x^2 h(x) \end{bmatrix} = \begin{bmatrix} 1&1&1&0&1&0&0 \\ 0&1&1&1&0&1&0 \\ 0&0&1&1&1&0&1 \end{bmatrix}.$$

Writing the columns of G' last to first, a generator matrix for $C^\perp$ is

$$H = \begin{bmatrix} 0&0&1&0&1&1&1 \\ 0&1&0&1&1&1&0 \\ 1&0&1&1&1&0&0 \end{bmatrix}.$$

A simple check verifies that $GH^T = 0$.

Since C' and $C^\perp$ can be obtained from each other by simply reversing the components in all vectors, C' and $C^\perp$ are equivalent codes. C' is often referred to as the dual code of C even though the dual is actually $C^\perp$. $h(x)$ is often said to generate the dual code even though it actually generates a code which is only equivalent to the dual. It is easy to see from the construction of H that $C^{\perp}$ is also a cyclic code.

In order to describe $C^{\perp}$ in terms of a generator polynomial for a cyclic code C, it is convenient to make the following definition.

Definition. Let $h(x)=\sum_{i=0}^{k} a_i x^i$ be a polynomial of degree k ($a_k \neq 0$). Define the *reciprocal polynomial* $h_R(x)$ of $h(x)$ by

$$h_R(x)=\sum_{i=0}^{k} a_{k-i} x^i.$$

Note that $h_R(x)=x^k \cdot h(1/x)$, where $k=\deg h(x)$. With this, we then have the following result.

Theorem 5.10. Let $g(x)$ be a monic divisor of $f(x)=x^n-1$ (over F) of degree $n-k$, and hence the generator for a cyclic (n,k)-code. Let $h(x)=f(x)/g(x)$. Then $h_R(x)$, the reciprocal polynomial of $h(x)$, generates $C^{\perp}$.

The polynomial $h(x)$ itself is sometimes called the *parity-check polynomial.*

5.5 Encoding Cyclic Codes†

Let $g(x)$ be the generating polynomial for a cyclic (n,k)-code C over F. A generator matrix G for C of the form $[R\ I_k]$ may be constructed as follows. Divide x^{n-k+i} by $g(x)$ for $0 \le i \le k-1$. This gives

$$x^{n-k+i}=q_i(x)g(x)+r_i(x)$$

where $\deg r_i(x) < \deg g(x) = n-k$ or $r_i(x)=0$. Then

$$x^{n-k+i}-r_i(x)=q_i(x)g(x) \in C.$$

† This section may be omitted without loss of continuity.

If we take the coefficient vectors corresponding to $x^{n-k+i} - r_i(x)$ as the rows of a matrix, we get a generator matrix of the form $[R\ I_k]$, where the rows of R correspond to $-r_i(x)$, $0 \le i \le k-1$.

Example 11.

Consider the binary cyclic (7,4)-code of example 10, generated by $g(x) = 1 + x + x^3$. By the division algorithm, we compute

$$x^3 = (1)(x^3+x+1) + (1+x)$$

$$x^4 = (x)(x^3+x+1) + (x+x^2)$$

$$x^5 = (x^2+1)(x^3+x+1) + (1+x+x^2)$$

$$x^6 = (x^3+x+1)(x^3+x+1) + (1+x^2).$$

A generator matrix for the code is hence

$$G = \begin{bmatrix} 1&1&0&1&0&0&0 \\ 0&1&1&0&1&0&0 \\ 1&1&1&0&0&1&0 \\ 1&0&1&0&0&0&1 \end{bmatrix} = [R\ I_4] \quad \text{where } R = \begin{bmatrix} 1&1&0 \\ 0&1&1 \\ 1&1&1 \\ 1&0&1 \end{bmatrix}.$$

The rows of R correspond to the polynomials $1+x$, $x+x^2$, $1+x+x^2$ and $1+x^2$. A message $\mathbf{m}$=(1011) would be encoded as $\mathbf{c} = \mathbf{m}G = (100\ 1011)$.

Now consider the message polynomial

$$a(x) = \sum_{i=0}^{k-1} a_i x^i.$$

We know that $a(x)g(x)$ is a codeword but given $a(x)g(x)$ the information symbols are not evident. If $G = [R\ I_k]$ is a generator matrix for C, then the message symbols $(a_0 a_1 \dots a_{k-1})$ appear in the last k positions of the codeword. The codeword corresponding to the message $a(x)$ can be determined through polynomial operations as follows. Divide $x^{n-k}a(x)$ by $g(x)$ to get

$$x^{n-k}a(x) = q(x)g(x) + t(x)$$

or

$$q(x)g(x) = -t(x) + x^{n-k}a(x),$$

where $t(x) = 0$ or $\deg t(x) < \deg g(x) = n - k$. Then $q(x)g(x)$ is a codeword with the k information symbols $(a_0a_1...a_{k-1})$ as the last k components, and $n-k$ parity-check symbols, the coefficients of $-t(x)$, as the first $n-k$ components.

Example 12.

Again, consider $g(x) = 1 + x + x^3$, which generates a binary cyclic (7,4)-code. Suppose we wish to find a codeword **c** which has the information symbols (1011) in the last 4 positions. Let $a(x) = 1 + x^2 + x^3$. Dividing $x^3a(x)$ by $g(x)$,

$$x^3a(x) = (x^3+x^2+x+1)g(x) + 1$$

and

$$(x^3+x^2+x+1)g(x) = 1 + x^3 + x^5 + x^6.$$

Therefore **c** = (100 1011) is the codeword with the information symbols (1011) in the high order positions. Compare to example 11.

Let C be a cyclic (n,k)-code with generator polynomial

$$g(x) = g_0 + g_1x + ... + g_{n-k-1}x^{n-k-1} + x^{n-k}$$

and generator matrix $G = [R\ I_k]$. The rows of R are the negated coefficients of the remainder polynomials $r_i(x)$, $0 \le i \le k-1$ where

$$x^{n-k+i} = q_i(x)g(x) + r_i(x),$$

$\deg r_i(x) < \deg g(x)$ or $r_i(x) = 0$. The $n-k$ parity-check symbols for the information symbols $(a_0a_1...a_{k-1})$ are given by

$$\sum_{i=0}^{k-1} a_ir_i(x).$$

In the binary case we can translate these observations into a simple efficient algorithm, suitable for hardware implementation, for encoding the message $\mathbf{m} = (a_0a_1...a_{k-1})$ to the codeword $\mathbf{c} = (s_0s_1...s_{n-k-1}a_0a_1...a_{k-1})$.

Encoding Algorithm for Binary Cyclic Codes.

Let $\hat{\mathbf{g}} = (g_0 g_1 \cdots g_{n-k-1})$.

Let the message symbols be $(a_0 a_1 \cdots a_{k-1})$.

Let the parity-check symbols to be found be $\mathbf{s} = (s_0 s_1 \cdots s_{n-k-1})$.

(1) Set $s_j = 0,\ 0 \le j \le n-k-1$.

(2) Set $i = 1$.

(3) If $a_{k-i} = s_{n-k-1}$ then

for j from $n-k-1$ to 1, set $s_j = s_{j-1}$.

$s_0 = 0$.

Otherwise

for j from $n-k-1$ to 1, set $s_j = s_{j-1} + g_j$.

$s_0 = g_0$.

(4) $i = i + 1$.

(5) If $i > k$, stop. Otherwise, go to (3).

A brief explanation of the algorithm follows. Using the generator matrix $G = [R\ I_k]$, then as discussed above, the $n-k$ parity-check symbols corresponding to the message $a(x)$ are the coefficients of $-t(x)$ where

$$x^{n-k} a(x) = q(x) g(x) + t(x).$$

To compute $t(x)$, we compute $x^{n-k} a(x) \pmod{g(x)}$. Now since

$$a(x) = a_0 + a_1 x + \ldots + a_{k-1} x^{k-1},$$

we have

$$x^{n-k} a(x) = a_0 x^{n-k} + a_1 x^{n-k+1} + \ldots + a_{k-1} x^{n-1}.$$

Nesting the multiplications, this may be computed as

$$(\ldots (((a_{k-1} x^{n-k}) x + a_{k-2} x^{n-k}) x + a_{k-3} x^{n-k}) x + \ \ldots\ ) x + a_0 x^{n-k}.$$

Each parenthesized expression, starting with the innermost, corresponds to an iteration of the algorithm. Reduction modulo $g(x)$ is carried out during each iteration, and the current partial sum, which corresponds to s, is always a polynomial of degree at most $n-k-1$. During the ith

iteration, s is multiplied by x (corresponding to a logical shift of s), and to this is added $a_{k-i}x^{n-k}$. Now s may have degree $n-k$, due to the appearance of the term x^{n-k} from either the shift or the addition. To carry out reduction modulo $g(x)$, the term x^{n-k} is reduced by deleting it and adding back $\hat{g}(x) = g(x) - x^{n-k}$ (since $x^{n-k} \equiv \hat{g}(x) \pmod{g(x)}$). This is sometimes referred to as "using feedback". If $s_{n-k-1} = 1$ before the shift of s, then the shift will produce an x^{n-k} term, necessitating feedback. If $a_{k-i} = 1$, this also necessitates feedback. If both s_{n-k-1} and a_{k-i} are 1, then $x^{n-k} + x^{n-k} \equiv 0 \pmod 2$, and they cancel one another. Hence feedback is necessary during iteration i if and only if $s_{n-k-1} \neq a_{k-i}$. After the kth iteration, the $(n-k)$-tuple s corresponds to $x^{n-k}a(x) \pmod{g(x)}$, and provides the $n-k$ parity-check symbols sought.

Note that using this encoding scheme, the generator matrix $G = [R\ I_k]$ need not be constructed explicitly or stored.

Example 13.

In the previous example, $g(x) = 1 + x + x^3$ generates a binary cyclic (7,4)-code. We encode the message $\mathbf{m} = (1011)$ as follows. We have $\hat{\mathbf{g}} = (g_0 g_1 g_2) = (110)$, $(a_0 a_1 a_2 a_3) = (1011)$, and start with $(s_0 s_1 s_2) = (000)$.

i	s	a_{k-i}
0	000	
1	110	1
2	101	1
3	100	0
4	100	1

The parity-check symbols are (100) and we encode to the codeword c=(100 1011).

Example 14.

Let $g(x) = 1 + x^4 + x^6 + x^7 + x^8$ be the generator for a binary (15,7)-code C. We encode the message $\mathbf{m} = (1011011)$ as follows. We have $\hat{\mathbf{g}} = (1000\ 1011)$.

i	s	a_{k-i}
0	0000 0000	
1	1000 1011	1
2	0100 0101	1
3	1010 1001	0
4	0101 0100	1
5	1010 0001	1
6	1101 1011	0
7	0110 1101	1

The message **m** is encoded to the codeword **c** = (0110 1101 1011 011).

5.6 Syndromes and Simple Decoding Procedures

Suppose C is a cyclic (n,k)-code over F and let $g(x)$ be the generator polynomial for C. From §5.5, a generator matrix is

$$G = [R\ I_k]$$

where R is a $k \times n-k$ matrix whose i^{th} row is derived from the coefficients of $r_i(x) = x^{n-k+i} - q_i(x)g(x), 0 \le i \le k-1$. We write

$$R = \begin{bmatrix} -r_0(x) \\ -r_1(x) \\ \cdot \\ \cdot \\ -r_{k-1}(x) \end{bmatrix}.$$

A parity-check matrix for C is

$$H = [I_{n-k}\ -R^T].$$

If $\mathbf{r}$ is an n-tuple, then $H\mathbf{r}^T = \mathbf{s}$ is the syndrome of $\mathbf{r}$. Of course, the syndrome of a vector depends on the parity-check matrix. One reason for this particular choice of H as a parity-check matrix is the convenient polynomial interpretation that it provides for the syndrome $\mathbf{s} = H\mathbf{r}^T$ of a vector $\mathbf{r}$, as given by the following result.

Theorem 5.11. Let $r(x)$ and $s(x)$ be the respective polynomial representations of a vector **r** and its syndrome **s**. Then $s(x)$ is the remainder polynomial when $r(x)$ is divided by $g(x)$.

Proof.

Suppose $r(x) = a_0 + a_1x + \cdots + a_{n-1}x^{n-1}$. Noting that column i, $0 \le i \le n-k-1$, of H corresponds to the polynomial x^i and column i, $n-k \le i \le n-1$, corresponds to $r_{i-n+k}(x)$,

$\mathbf{s} = H\mathbf{r}^T$, so that

$$\begin{aligned} s(x) &= a_0 + a_1x + \cdots + a_{n-k-1}x^{n-k-1} + a_{n-k}r_0(x) + \cdots + a_{n-1}r_{k-1}(x) \\ &= a_0 + a_1x + \cdots + a_{n-k-1}x^{n-k-1} \\ &\quad + a_{n-k}(x^{n-k} - q_0(x)g(x)) + \cdots + a_{n-1}(x^{n-1} - q_{k-1}g(x)) \\ &= r(x) - (\, a_{n-k}q_0(x)g(x) + \cdots + a_{n-1}q_{k-1}(x)g(x)\,) \\ &= r(x) - h(x)g(x) \end{aligned}$$

where $h(x) = a_{n-k}q_0(x) + \cdots + a_{n-1}q_{k-1}(x)$. Thus $s(x) = r(x) - h(x)g(x)$, which implies $r(x) = h(x)g(x) + s(x)$. Now since $\deg r_i(x) \le n-k-1$, it follows that $\deg s(x) \le n-k-1$. Hence by the uniqueness of the quotient and remainder in the division algorithm for polynomials, we conclude that $s(x)$ is the remainder when $r(x)$ is divided by $g(x)$. □

Therefore the syndrome of a vector can be determined by polynomial division. Furthermore, the syndrome of a vector and its cyclic shift are closely related. Suppose

$$r(x) = q(x)g(x) + s(x),$$

where $s(x)$ has degree at most $n-k-1$. Then by Theorem 5.11, $s(x)$ is the syndrome of $r(x)$. Now

$$xr(x) = xq(x)g(x) + xs(x).$$

If $xs(x)$ has degree at most $n-k-1$, then again by Theorem 5.11, it is the syndrome of $xr(x)$. Otherwise, the syndrome of $xr(x)$ is the remainder when $xs(x)$ is divided by $g(x)$. In this case, let

$$s(x) = \sum_{i=0}^{n-k-1} s_i x^i = \hat{s}(x) + s_{n-k-1}x^{n-k-1}, \quad \deg \hat{s}(x) \le n-k-2.$$

Let the (monic) generator polynomial be

$$g(x) = \sum_{i=0}^{n-k} g_i x^i = \hat{g}(x) + x^{n-k}, \quad \deg \hat{g}(x) \le n-k-1.$$

Then

$$xs(x) = x\hat{s}(x) + s_{n-k-1}(g(x) - \hat{g}(x))$$
$$= s_{n-k-1}g(x) + (x\hat{s}(x) - s_{n-k-1}\hat{g}(x)),$$

where $\deg(x\hat{s}(x) - s_{n-k-1}\hat{g}(x)) \le n-k-1$. Now by the uniqueness of the remainder in the division algorithm, and Theorem 5.11, we see that the syndrome of $xs(x)$ is $xs(x) - s_{n-k-1}g(x)$. We summarize as follows.

Theorem 5.12. Let C be a cyclic (n,k)-code over F with generator polynomial $g(x)$, and let $r(x)$ be a polynomial with syndrome $s(x) = \sum_{i=0}^{n-k-1} s_i x^i$. Then the syndrome of $x\,r(x)$ is

(1) $x\,s(x)$, if $\deg s(x) < n-k-1$, and

(2) $x\,s(x) - s_{n-k-1}g(x)$, if $\deg s(x) = n-k-1$.

Example 15.

Let $g(x) = 1 + x + x^3$ be the generator polynomial for a binary cyclic (7,4)-code. From example 11, a generator matrix for the code is

$$G = [R\ I_4] = \begin{bmatrix} 1&1&0&1&0&0&0 \\ 0&1&1&0&1&0&0 \\ 1&1&1&0&0&1&0 \\ 1&0&1&0&0&0&1 \end{bmatrix}.$$

A parity-check matrix is therefore

$$H = [I_3 - R^T] = \begin{bmatrix} 1&0&0&1&0&1&1 \\ 0&1&0&1&1&1&0 \\ 0&0&1&0&1&1&1 \end{bmatrix}.$$

Consider $\mathbf{r} = (1011011)$. The syndrome of $\mathbf{r}$ is $H\mathbf{r}^T = (001)^T = \mathbf{s}$. Now

$$r(x) = 1 + x^2 + x^3 + x^5 + x^6.$$

If we divide $r(x)$ by $g(x)$, we get

$$r(x) = (x^3+x^2+x+1)g(x) + x^2.$$

The remainder polynomial is $s(x) = x^2$, and as expected this is the syndrome which we obtain from the computation $H\mathbf{r}^T$.

Let us now compute the syndrome of a cyclic shift of $\mathbf{r}$. A cyclic shift of $\mathbf{r}$ yields the vector $\mathbf{w} = (1101101)$, corresponding to the polynomial $xr(x)$. Now

$$H\mathbf{w}^T = (110)^T = \mathbf{t}.$$

We can compute this syndrome using polynomial operations as follows. Since $\deg s(x) = 2 = n-k-1$, $t(x)$ is

$$xs(x) - 1\cdot g(x) = x^3 - (x^3+x+1) = 1 + x.$$

The preceding observations lead us to an interesting method of decoding certain error patterns in cyclic codes. The technique is sometimes referred to as *error trapping*. We first make the following definition.

Definition. A *cyclic run* of length $k \le n$ is a succession of k cyclically consecutive components in an n-tuple.

For example, $\mathbf{e} = (0100\ 0101)$ has a cyclic run of three 0's, and $\mathbf{e} = (0011\ 0100)$ has a cyclic run of four 0's.

We now proceed to describe a decoding strategy. Suppose C is an (n,k)-code with distance $d = 2t + 1$ and parity-check matrix $H = [I_{n-k}\ A]$. Let $\mathbf{c}$ be a codeword and $\mathbf{e}$ be an error pattern of weight at most t. If $\mathbf{r} = \mathbf{c} + \mathbf{e}$ is received, then the syndrome of $\mathbf{r}$ is

$$\mathbf{s} = H\mathbf{r}^T = H(\mathbf{c}+\mathbf{e})^T = H\mathbf{e}^T.$$

Let $\hat{\mathbf{e}} = (\mathbf{s}^T, \mathbf{0})$, where $\mathbf{0}$ is the k-tuple of all 0's. Then $H\hat{\mathbf{e}}^T = \mathbf{s}$, and $\hat{\mathbf{e}}$ and $\mathbf{e}$ have the same syndrome, implying by Theorem 3.10 that $\hat{\mathbf{e}}$ and $\mathbf{e}$ are in the same coset of C. Now suppose $w(\mathbf{s}) \le t$. Then $w(\hat{\mathbf{e}}) \le t$, and it follows that $\mathbf{e} = \hat{\mathbf{e}}$, since by Theorem 3.9, any coset of C contains at most one vector of weight less than or equal to t. Hence the error is known to be $\mathbf{e} = (\mathbf{s}^T, \mathbf{0})$.

Now suppose C is cyclic with generating polynomial $g(x)$. Let $\mathbf{e}$ be an error pattern of weight at most t with a cyclic run of at least k 0's. Then there is some i, $0 \le i \le n-1$, such that a cyclic shift of $\mathbf{e}$ through i positions is a vector whose non-zero components all lie within the first $n-k$ components. For this i, $w(s_i(x)) \le t$, where $s_i(x)$ is the syndrome of $x^i e(x) \pmod{x^n-1}$. If we compute $s_i(x)$ as the remainder of $x^i r(x)$ divided by $g(x)$, then when $w(s_i(x)) \le t$, for this i, by the above argument,

$$x^i e(x) = (\mathbf{s}_i, \mathbf{0}).$$

Thus $e(x) = x^{n-i}(\mathbf{s}_i,\mathbf{0})$, where $x^{n-i}(\mathbf{s}_i,\mathbf{0})$ is a cyclic shift of vector $(\mathbf{s}_i,\mathbf{0})$ through $n-i$ positions (i.e. $x^{n-i}s_i(x) \pmod{x^n - 1}$). Computing $s_i(x)$ from $s_{i-1}(x)$ using Theorem 5.12, this gives rise to the following algorithm.

Decoding Algorithm for Binary Cyclic Codes (Error Trapping).

Let C be a t-error correcting cyclic (n,k)-code with generator polynomial $g(x)$.
Let $s_i(x)$ denote the syndrome of $x^i r(x)$.
Let $e(x)$ be an error pattern, $w(\mathbf{e}) \le t$, containing a cyclic run of at least k 0's.
Decode the received vector $r(x) = c(x) + e(x)$ to $c(x)$ as follows.

(1) Compute the syndrome $s_0(x)$ of $r(x)$ from the division algorithm.

$$r(x) = q(x)g(x) + s_0(x).$$

(2) $i = 0$.

(3) If $w(s_i(x)) \le t$, then set $e(x) = x^{n-i}(\mathbf{s}_i,\mathbf{0})$, and decode to $r(x) - e(x)$.

(4) Set $i = i+1$.

(5) If $i = n$ then stop; the error pattern is not trappable.

(6) If $\deg s_{i-1}(x) < n-k-1$, then set $s_i(x) = x\, s_{i-1}(x)$;
otherwise set $s_i(x) = x\, s_{i-1}(x) - g(x)$.

(7) Go to (3).

To use the above algorithm for non-binary cyclic codes, the computation of $s_i(x)$ in step (6) should be replaced by the more general form of Theorem 5.12.

Example 16.

$g(x) = 1 + x^2 + x^3$ generates a binary (7,4)-cyclic code C having distance 3. Consider the code polynomial $c(x) = a(x)g(x)$ where $a(x) = 1 + x + x^2$. That is, $c(x) = 1 + x + x^5$. Suppose on transmission the error polynomial $e(x) = x^6$ is introduced. Then $r(x) = 1 + x + x^5 + x^6$ is received at the decoder. We compute the syndrome of $r(x)$ by dividing by $g(x)$.

$$r(x) = (x^3+1)g(x) + (x+x^2)$$

$$s(x) = x + x^2.$$

Since $w(s(x)) > 1$ we compute the syndrome $s_1(x)$ of $xr(x)$. This is easily obtained from $s(x)$. Since $\deg s(x) = 2 = n-k-1$, we multiply $s(x)$ by x and subtract $g(x)$ to get $s_1(x) = 1$. Since $w(s_1) \le 1$ we determine the error pattern to be

$$e(x) = x^{7-1}(\mathbf{s}_1, 0) = x^6(1000000) = x^6.$$

Since for n=7 all error patterns of weight one have a cyclic run of six 0's, and $6 \ge k = 4$, this technique will correct all single errors in C.

Example 17.

$g(x) = 1 + x^4 + x^6 + x^7 + x^8$ generates a (15,7)-cyclic code over Z_2. Suppose the distance in this code is known to be 5. Any error pattern having weight at most 2 must contain a run of 0's of length at least 7. Thus, the decoding procedure outlined above will correct all single and double errors. Using this technique we decode the received 15-tuple

$$\mathbf{r} = (1100\ 1110\ 1100\ 010).$$

We first compute the syndrome polynomial $s(x)$ of $r(x)$ by the division algorithm.

$$r(x) = (x^5+x^4+x^2+x)g(x) + (1 + x^2 + x^5 + x^7)$$

$$s(x) = 1 + x^2 + x^5 + x^7.$$

We compute the syndrome $s_i(x)$ of $x^i r(x)$ until $w(s_i(x)) \le 2$.

i	$s_i(x)$
0	1010 0101
1	1101 1001
2	1110 0111
3	1111 1000
4	0111 1100
5	0011 1110
6	0001 1111
7	1000 0100

Since $w(\mathbf{s}_7) \le 2$ the error pattern is

$$\mathbf{e} = x^{15-7}(\mathbf{s}_7,0) = x^8(1000\ 0100\ 0000\ 000) = (0000\ 0000\ 1000\ 010).$$

We decode $\mathbf{r}$ to $\mathbf{r}-\mathbf{e} = (1100\ 1110\ 0100\ 000)$.

Example 18.

$g(x) = 1+x+x^2+x^4+x^5+x^8+x^{10}$ generates a cyclic (15,5)-code having distance 7. Any error pattern of weight 3 or less must contain a run of at least 5 0's unless it is $\hat{\mathbf{e}} = (10000\ 10000\ 10000)$ or any cyclic shift of this. Thus, the decoding scheme of this section will correct all error patterns of weight 3 or less except for $\hat{\mathbf{e}}$ and its cyclic shifts. We can modify our decoding scheme to correct these error patterns also. Note that the syndrome of $\hat{e}(x)$ is $1+x^5+r_1(x)$ where $r_1(x)$ is the remainder when x^{10} is divided by $g(x)$. So for a received vector $\mathbf{r}$ we compute the syndromes $s_i(x)$ of $x^i r(x)$, $0 \le i \le 14$, and now check whether $w(\mathbf{s}_i) \le 3$ or $w(\mathbf{s}_i - \mathbf{r}_1) \le 2$. In the latter case, the error pattern is $x^{15-i}(\mathbf{s}_i - \mathbf{r}_1, (10000))$.

(i) Suppose the decoder receives

$$\mathbf{r} = (11110\ 10100\ 11101).$$

We compute $s_i(x)$ starting at $i = 0$.

i	$s_i(x)$
0	01100 00100

Since $w(\mathbf{s}_0) \le 3$, the error pattern is

$$\mathbf{e} = x^{15-0}(\mathbf{s}_0,0) = (01100\ 00100\ 00000).$$

We decode $\mathbf{r}$ to $\mathbf{r}-\mathbf{e} = (10010\ 10000\ 11101)$.

(ii) Suppose the decoder receives

$$\mathbf{r} = (11100\ 01111\ 00100).$$

We compute the syndromes $s_i(x)$ of $x^i r(x)$, $i = 0, 1, 2, \ldots.$ We also compute $s_i(x) - r_1(x)$, where $r_1(x)$ is found to be $r_1(x) = 1+x+x^2+x^4+x^5+x^8$.

i	$s_i(x)$	$s_i(x)-r_1(x)$
0	00110 10001	11011 00011
1	11110 11010	00011 01000
2	01111 01101	10010 11111
3	11010 00100	00111 10110
4	01101 00010	10000 10000

Since $w(\mathbf{s}_4-\mathbf{r}_1) \leq 2$ the error pattern is

$$\mathbf{e} = x^{11}(10000\ 10000\ 10000) = (01000\ 01000\ 01000).$$

We decode $\mathbf{r}$ to $\mathbf{r} - \mathbf{e} = (10100\ 00111\ 01100)$.

5.7 Burst Error Correcting

The codes we have considered so far have been designed to correct random errors. In general, a t-error correcting code corrects all error patterns of weight t or less in a codeword of block length n. It may be, however, that certain channels introduce errors localized in short intervals rather than at random. For example, in storage mediums, errors resulting from physical irregularities or structural alteration, perhaps flaws in the original medium or damage due to wear and tear, are not independent, but rather tend to be spatially concentrated. Similarly, interference over short time intervals in serially transmitted radio signals causes errors to occur in *bursts*. There exist codes for correcting such *burst errors*. Many of these codes are cyclic. We briefly consider burst-error correcting codes in this section.

Definition. A *cyclic burst error* of length t is a vector whose non-zero components are contained within a cyclic run of length t, the first and last components in the run being non-zero.

Note that a cyclic burst may wrap around from the end of the vector to the beginning.

Example 19.

(i) $\mathbf{e}_1 = (0101\ 0110\ 000)$ is a burst of length 6 in $V_{11}(Z_2)$.

(ii) $\mathbf{e}_2 = (0000\ 0010\ 001)$ is a burst of length 5 in $V_{11}(Z_2)$.

(iii) $\mathbf{e}_3 = (0100\ 0000\ 100)$ is a burst of length 5 in $V_{11}(Z_2)$.

We can describe a burst error of length t in terms of a polynomial as

$$e(x) = x^i b(x) \pmod{x^n - 1},$$

where $b(x)$ is a polynomial of degree $t-1$ which describes the error pattern, and i indicates where the burst begins. For the examples above, the polynomial representations of the burst errors are

$$e_1(x) = x(1+x^2+x^4+x^5),$$

$$e_2(x) = x^6(1+x^4), \text{ and}$$

$$e_3(x) = x^8(1+x^4).$$

Consider a linear code C. If all burst errors of length t or less occur in distinct cosets of a standard array for C, then each can be uniquely identified by its syndrome, and all such errors are then correctable. Furthermore, if C is a linear code capable of correcting all burst errors of length t or less, then all such errors must occur in distinct cosets. To see this, suppose C can correct two such distinct errors $\mathbf{b}_1$ and $\mathbf{b}_2$ which lie in some coset C_i of C. Then $\mathbf{b}_1 - \mathbf{b}_2 = \mathbf{c}$ is a non-zero codeword. Now suppose $\mathbf{b}_1$ is a received vector. How should it be decoded? The codeword $\mathbf{0}$ could have been altered to $\mathbf{b}_1$ by the error $\mathbf{b}_1$, and the codeword $\mathbf{c}$ could have been altered to $\mathbf{b}_1$ by the error $\mathbf{b}_2$. We get a contradiction, since the code cannot correct both errors $\mathbf{b}_1$ and $\mathbf{b}_2$. We conclude that $\mathbf{b}_1$ and $\mathbf{b}_2$ must lie in distinct cosets. This gives us the following result.

Theorem 5.13. A linear code C can correct all burst errors of length t or less if and only if all such errors occur in distinct cosets of C.

The result is true not only for burst errors of length t, but for any set of errors one might wish to correct. In general, for any codewords $\mathbf{c}_i$ and $\mathbf{c}_j$, if $\mathbf{c}_i + \mathbf{e}_1 = \mathbf{c}_j + \mathbf{e}_2$, then $\mathbf{e}_1$ and $\mathbf{e}_2$ are not both correctable errors.

It follows that a cyclic code can correct all burst errors of length t or less if and only if the syndrome polynomials for these bursts are distinct.

Example 20.

A (15,9)-cyclic code C is generated by $g(x) = 1 + x + x^2 + x^3 + x^6$. C can correct all cyclic bursts of length 3 or less. We check this by computing the syndrome of each burst error pattern. The bursts of weight 1 are

$$e(x)=x^i, 0 \le i \le 14.$$

The syndrome of each of these bursts is given by the following table. To facilitate checking for duplicate syndromes, we include with each syndrome the binary integer corresponding to the binary interpretation of that syndrome.

Burst Error	Syndrome	Binary Value
x^0	100000	1
x^1	010000	2
x^2	001000	4
x^3	000100	8
x^4	000010	16
x^5	000001	32
x^6	111100	15
x^7	011110	30
x^8	001111	60
x^9	111011	55
x^{10}	100001	33
x^{11}	101100	13
x^{12}	010110	26
x^{13}	001011	52
x^{14}	111001	39

The bursts of length 2 and their syndromes are as follows.

Burst Error	Syndrome	Binary Value
$1+x$	110000	3
$x(1+x)$	011000	6
$x^2(1+x)$	001100	12
$x^3(1+x)$	000110	24
$x^4(1+x)$	000011	48
$x^5(1+x)$	111101	47
$x^6(1+x)$	100010	17
$x^7(1+x)$	010001	34
$x^8(1+x)$	110100	11
$x^9(1+x)$	011010	22
$x^{10}(1+x)$	001101	44
$x^{11}(1+x)$	111010	23
$x^{12}(1+x)$	011101	46
$x^{13}(1+x)$	110010	19
$x^{14}(1+x)$	011001	38

We see that all syndromes so far are distinct, as none of the binary values has been duplicated. The bursts of length 3 are given by the expressions

$$e(x) = x^i(1+x^2), \quad 0 \le i \le 14$$

$$e(x) = x^i(1+x+x^2), \quad 0 \le i \le 14.$$

These bursts and their syndromes are given in the following tables.

Burst Error	Syndrome	Binary Value
$1+x^2$	101000	5
$x(1+x^2)$	010100	10
$x^2(1+x^2)$	001010	20
$x^3(1+x^2)$	000101	40
$x^4(1+x^2)$	111110	31
$x^5(1+x^2)$	011111	62
$x^6(1+x^2)$	110011	51
$x^7(1+x^2)$	100101	41
$x^8(1+x^2)$	101110	29
$x^9(1+x^2)$	010111	58
$x^{10}(1+x^2)$	110111	59
$x^{11}(1+x^2)$	100111	57
$x^{12}(1+x^2)$	101111	61
$x^{13}(1+x^2)$	101011	53
$x^{14}(1+x^2)$	101001	37

Burst Error	Syndrome	Binary Value
$1+x+x^2$	111000	7
$x(1+x+x^2)$	011100	14
$x^2(1+x+x^2)$	001110	28
$x^3(1+x+x^2)$	000111	56
$x^4(1+x+x^2)$	111111	63
$x^5(1+x+x^2)$	100011	49
$x^6(1+x+x^2)$	101101	45
$x^7(1+x+x^2)$	101010	21
$x^8(1+x+x^2)$	010101	42
$x^9(1+x+x^2)$	110110	27
$x^{10}(1+x+x^2)$	011011	54
$x^{11}(1+x+x^2)$	110001	35
$x^{12}(1+x+x^2)$	100100	9
$x^{13}(1+x+x^2)$	010010	18
$x^{14}(1+x+x^2)$	001001	36

There are no other burst error patterns of length 3. The 60 syndromes corresponding to the 60 distinct burst errors of length 3 or less given above are all distinct. Hence we conclude that C can correct all bursts of length 3 or less. Note that since there are only 63 non-zero binary 6-tuples, this condition is not easily satisfied.

We can decode this particular code by the error-trapping method of the previous section. Recall that the syndrome obtained from polynomial division is the syndrome obtained from the parity-check matrix $H = [I_{n-k}\ -R^T]$ where R^T contains remainder polynomials as columns. If **e** is a burst error of burst length at most 3 and the burst occurs in the first 6 components of the vector, then $H\mathbf{e}^T = \mathbf{s}$ is a (non-cyclic) burst of length at most 3 which describes the error positions. If the errors do not occur in the first 6 positions, then as before some cyclic shift of **e** will give a syndrome which is a burst of length at most 3, and from this syndrome, **e** can be determined.

In fact, error-trapping can be used for any cyclic burst-error correcting code. It can be proven that a t-burst-error correcting (n,k)-code satisfies $n-k \geq 2t$ (exercise 22). Hence $n-k \geq t$ and $n-t \geq k$. Now a burst error of length t in a codeword of length n has a cyclic run of $n-t$ 0's, implying that every burst error of length t or less has a cyclic run of at least k 0's, which is the requirement of the error-trapping algorithm. Hence we have the following modification of the error-trapping algorithm, which can be used to trap all bursts of length t or less in any cyclic t-burst-error correcting code.

Error-trapping for Cyclic Burst-Error Codes.

Let C be a code capable of correcting all burst errors of length t or less. Denote by $s_i(x)$ the syndrome of $x^i r(x)$.

Let $r(x)$ be a received vector.

(1) By the division algorithm, compute the syndrome $s_0(x)$ of $r(x)$.

(2) $i = 0$.

(3) If $s_i(x)$ is a (non-cyclic) burst of length t or less, then $e(x) = x^{n-i}(\mathbf{s}_i, \mathbf{0})$.

(4) $i = i + 1$.

(5) If $i = n$, then stop; the error pattern is not trappable.

(6) Compute $s_i(x)$ (as in the original algorithm).

(7) Go to (3).

Example 21.

The code C of example 20 is 3-burst-error correcting and has generator polynomial $g(x) = 1 + x + x^2 + x^3 + x^6$. We correct the received vector $\mathbf{r} = (1110\ 1110\ 1100\ 000)$ as follows.

$$r(x) = (x^3+x^2)g(x) + 1 + x + x^4 + x^5.$$

$$s_0(x) = 1 + x + x^4 + x^5.$$

Since $w(\mathbf{s}_0) > 3$ we compute $s_1(x) = 1 + x^3 + x^5$ and proceed. We list the syndromes in the following table.

i	$s_i(x)$
0	110011
1	100101
2	101110
3	010111
4	110111
5	100111
6	101111
7	101011
8	101001
9	101000

Since $s_9(x)$ is a burst of length 3 we determine the error pattern as $\mathbf{e} = (0000\ 0010\ 1000\ 000)$. We decode $\mathbf{r}$ to

$$\mathbf{r} - \mathbf{e} = (1110\ 1100\ 0100\ 000).$$

Note that $s_8(x)$ is a syndrome of weight 3, but not a burst of length 3 or less.

As one might imagine from the computations necessary in example 20, many of the better burst-error correcting codes have been found by computer search. The table below gives a few examples of binary cyclic codes with generator polynomial $g(x)$, capable of correcting all burst errors of length t or less, for some small values of t.

$g(x)$	(n,k)	Burst-correctability t
$1+x^2+x^3+x^4$	(7,3)	2
$1+x^2+x^4+x^5$	(15,10)	2
$1+x^4+x^5+x^6$	(31,25)	2
$1+x^3+x^4+x^5+x^6$	(15,9)	3
$1+x+x^2+x^3+x^6$	(15,9)	3

A very simple and effective technique for increasing the ability of a code to correct burst errors is known as *interleaving*. This technique is discussed in Chapter 7.

Analytic methods have also been used to find burst error codes. The analytically constructible class of codes known as the *Fire codes* have very high rate and can be used to provide excellent burst-error correcting capability.

5.8 Finite Fields and Factoring x^n-1 over GF(q)

Factoring x^n-1 over $GF(q)$ is extremely important in the study of cyclic codes. We devote this section to a discussion of this problem.

Let $GF(q)$ have characteristic p. If n and q are not coprime, then $n=\hat{n}p^s$ for some positive integer s, where $\gcd(\hat{n},q)=1$; then by Lemma 2.7, $x^n-1=(x^{\hat{n}}-1)^{p^s}$. Hence we shall assume that $\gcd(n,q)=1$.

Let m be the order of q modulo n, i.e. the smallest positive integer such that $q^m \equiv 1$ (mod n). Then $q^m-1=kn$ for some integer k. Consider the finite field $F=GF(q^m)$. From Lemma 2.4, we know that every element of F is a root of the polynomial $x^{q^m}-x$. Using the formula to sum a geometric series, note that

$$x^n+x^{2n}+\cdots+x^{kn}=\frac{x^n(x^{kn}-1)}{x^n-1}.$$

Hence x^n-1 divides $x^{kn}-1=x^{q^m-1}-1$. It follows that x^n-1 has all of its roots in $F=GF(q^m)$, and that $GF(q^m)$ is the splitting field of x^n-1 over $GF(q)$ (see § 6.1).

If $\gamma\in F$ is a root of x^n-1, then $\gamma^n=1$ and γ is called an nth *root of unity*. Suppose α is a primitive element in F, so that α has order q^m-1. Then $\alpha^k=\alpha^{(q^m-1)/n}$ has order n and is a root of x^n-1; α^k is called a *primitive* n^{th} root of unity, since $(\alpha^k)^n=1$ and $(\alpha^k)^j\neq 1$ for all positive $j<n$. We shall next require the following definition.

Definition. Given q and n, and a fixed integer i, $0\le i\le n-1$, the *cyclotomic coset* (of q modulo n) containing i is defined to be

$$C_i=\{i, iq, iq^2, ..., iq^{s-1}\}$$

where the elements of the set are taken mod n, and s is the smallest integer such that $iq^s\equiv i$ (mod n). We call $C=\{C_i: 0\le i\le n-1\}$ the *set of cyclotomic cosets* of q modulo n.

Example 22.

For $n = 9$ and $q = 2$,

$$C_1 = \{1,2,4,8,7,5\} = C_2 = C_4 = C_8 = C_7 = C_5$$

$$C_3 = \{3,6\} = C_6$$

$$C_0 = \{0\}$$

The set of cyclotomic cosets of 2 mod 9 is then $\mathrm{C} = \{C_0, C_1, C_3\}$.

Example 23.

Consider $n = 13$, $q = 3$. Then

$$C_1 = \{1,3,9\} = C_3 = C_9$$

$$C_2 = \{2,6,5\} = C_6 = C_5$$

$$C_4 = \{4,12,10\} = C_{10} = C_{12}$$

$$C_7 = \{7,8,11\} = C_8 = C_{11}$$

$$C_0 = \{0\}.$$

If we define a relation R on the integers by the rule $a\,R\,b$ if and only if for some integer j, $a \equiv bq^j \pmod n$, then for $\gcd(n,q) = 1$, it is easy to show that this relation is an equivalence relation on the integers modulo n, and the equivalence classes are the cyclotomic cosets of q mod n (exercise 49). Hence if $S, T \in \mathrm{C}$ and are distinct, then $S \cap T = \emptyset$ and $\bigcup_{S \in \mathrm{C}} S = Z_n$. We will shortly proceed to show that the number of irreducible factors of $f(x) = x^n - 1$ over $GF(q)$ is equal to the number of distinct cyclotomic cosets of q modulo n.

We first review some material regarding minimal polynomials, but now with respect to $GF(q)$ (as opposed to $GF(p)$, as in §2.4). The *minimal polynomial* of an element $\beta \in GF(q^m)$, *with respect to GF(q)*, is the monic polynomial $m(x) \in GF(q)[x]$ of smallest degree satisfying $m(\beta) = 0$. As in Theorem 2.8 and Theorem 2.9, this minimal polynomial is easily shown to be unique and irreducible (over $GF(q)$ now). For any $\beta \in GF(q^m)$, the *conjugates* of β, with respect to $GF(q)$, are the elements $\beta, \beta^q, \beta^{q^2}, \ldots, \beta^{q^{t-1}}$, where t is the smallest positive integer such that $\beta^{q^t} = \beta$. Analogous to the result of Theorem 2.11 then, such an element β has minimal polynomial with respect to $GF(q)$ being precisely

$$m_\beta(x) = \prod_{i=0}^{t-1} (x-\beta^{q^i}).$$

Let us now return to the problem of determining the factors of x^n-1 over $GF(q)$. Again, let α be a primitive element for $GF(q^m)$, and let $q^m-1=kn$, so that α^k is a primitive n^{th} root of unity. First, we note that x^n-1 has n distinct roots over $GF(q^m)$. This follows because x^n-1 and its derivative nx^{n-1} (which is non-zero, since $\gcd(n,q)=1$) have no factors in common (see exercise 50). Furthermore, these roots are precisely $(\alpha^k)^i$, $i=0,1,\ldots,n-1$. Now if $\beta=\alpha^{ki}$ is a root of x^n-1, then $\beta^n=1$ and each of the conjugates of β with respect to $GF(q)$, β^{q^j}, is also a root, since $(\beta^{q^j})^n=(\beta^n)^{q^j}=1$. Hence $m_\beta(x)$ is a factor of x^n-1. The roots of $m_\beta(x)$ are the elements

$$\alpha^{ki}, \alpha^{kiq}, \alpha^{kiq^2}, \ldots, \alpha^{kiq^{t-1}},$$

where t is the smallest positive integer such that $kiq^t \equiv ki \pmod{kn}$. This condition can be simplified to t being the smallest positive integer such that $iq^t \equiv i \pmod{n}$. It follows that the degree of $m_\beta(x)$ is the cardinality of the cyclotomic coset (of q mod n) containing i, $|C_i|$. Thus we can partition the set of roots of x^n-1 into $|C|$ classes, each class being the set of roots of an irreducible factor of x^n-1. We summarize these observations in the following theorem.

Theorem 5.14. Let $f(x)=x^n-1$ be a polynomial over $GF(q)$. The number of irreducible factors of $f(x)$ is equal to the number of cyclotomic cosets of q modulo n.

The procedure described above will, in fact, produce the factorization of x^n-1 but it requires that we do computations in the extension field $GF(q^m)$. We illustrate the method by example here, and describe a more convenient technique in §5.9.

Example 24.

Suppose we wish to factor $f(x)=x^{15}-1$ over $GF(2)$. Here n=15, q=2 and m=4. We first compute the cyclotomic cosets of 2 mod 15. These are

$C_0=\{0\}$

$C_1=\{1,2,4,8\}$

$C_3=\{3,6,12,9\}$

$$C_5 = \{5,10\}$$

$$C_7 = \{7,14,13,11\}.$$

This tells us that $x^{15}-1$ factors as a linear term, an irreducible quadratic and 3 irreducible quartics. One way to find these quartics is to proceed as in our earlier discussion. We will make use of the field $GF(2^4)$ generated using the polynomial $1+x+x^4$ (see Appendix D). If α is a primitive element of $GF(2^4)$, then α is a primitive 15th root of unity, and is a root of $x^{15}-1$. Hence α^2, α^4, and α^8 are also roots, and

$$m_\alpha(x) = (x-\alpha)(x-\alpha^2)(x-\alpha^4)(x-\alpha^8)$$

is a factor of $f(x)$. Expanding $m_\alpha(x)$ we get

$$m_\alpha(x) = x^4 + (\alpha+\alpha^2+\alpha^4+\alpha^8)x^3 + (\alpha^3+\alpha^5+\alpha^9+\alpha^6+\alpha^{10}+\alpha^{12})x^2$$

$$+ (\alpha^7+\alpha^{11}+\alpha^{13}+\alpha^{14})x + \alpha^{15}.$$

Using the Zech's log table it is easy to evaluate

$$\alpha+\alpha^2+\alpha^4+\alpha^8 = \alpha(1+\alpha)+\alpha^4(1+\alpha^4) = \alpha^5+\alpha^5 = 0,$$

$$\alpha^3+\alpha^5+\alpha^9+\alpha^6+\alpha^{10}+\alpha^{12} = \alpha^3(1+\alpha^6)+(\alpha^5+\alpha^{10})+\alpha^6(1+\alpha^6)$$

$$= \alpha+1+\alpha^4 = 0,$$

$$\alpha^7+\alpha^{11}+\alpha^{13}+\alpha^{14} = \alpha^7(1+\alpha^4)+\alpha^{13}(1+\alpha) = \alpha^8+\alpha^2 = 1.$$

Hence $m_\alpha(x) = x^4+x+1$ (as expected). In a similar manner, we can evaluate $m_{\alpha^3}(x)$, $m_{\alpha^5}(x)$ and $m_{\alpha^7}(x)$ to get

$$x^{15}-1 = (x-1)(x^4+x+1)(x^4+x^3+1)(x^2+x+1)(x^4+x^3+x^2+x+1).$$

As an alternative to finding the minimal polynomial of an element $\beta \in GF(q^m)$ by expanding and then simplifying the polynomial obtained as the product of linear factors corresponding to the conjugates of β, note that given the vector representations of the elements of $GF(q^m)$ (as included in the Zech's log tables in Appendix D – e.g. $\alpha^4 = (1100)$), a less arduous approach is to seek the coefficients of the first linear dependence over $GF(q)$ of the powers β^i of β ($i \le m$). For example, to determine $m_{\alpha^7}(x)$, note

$$(\alpha^7)^0 = (1000)$$

$$(\alpha^7)^1 = (1\,1\,0\,1)$$

$$(\alpha^7)^2 = (1\,0\,0\,1)$$

$$(\alpha^7)^3 = (0\,0\,1\,1)$$

$$(\alpha^7)^4 = (1\,0\,1\,1)$$

from which it can be determined that

$$1\cdot(\alpha^7)^0 + 1\cdot(\alpha^7)^3 + 1\cdot(\alpha^7)^4 = 0.$$

Hence $m_{\alpha^7}(x) = 1 + x^3 + x^4$.

Example 25.

We factor x^9-1 over $GF(2)$. The cyclotomic cosets of 2 modulo 9 are

$C_0 = \{0\}$

$C_1 = \{1,2,4,8,7,5\}$

$C_3 = \{3,6\}$.

Hence x^9-1 factors as a linear term, a quadratic term and an irreducible of degree 6. We observe that

$$\begin{aligned} x^9 - 1 &= (x^3)^3 - 1 = (x^3-1)(x^6+x^3+1) \\ &= (x-1)(x^2+x+1)(x^6+x^3+1). \end{aligned}$$

This must be the complete factorization.

Before proceeding with the next example, we make two observations regarding the reciprocal polynomial introduced in §5.4. Let $g(x) = \sum_{i=0}^{t} a_i x^i$ be a polynomial of degree exactly t in $F[x]$, and let $g_R(x) = x^t \cdot g(1/x)$. First, if α is a non-zero root of $g(x)$, then α^{-1} is a root of $g_R(x)$, since

$$g_R(\alpha^{-1}) = \sum_{i=0}^{t} a_{t-i}\alpha^{-i} = \alpha^{-t}\sum_{i=0}^{t} a_{t-i}\alpha^{t-i} = \alpha^{-t} g(\alpha) = 0.$$

Secondly, if $g(x)$ is irreducible, then so is $g_R(x)$. This follows since $g_R(x) = x^t \cdot g(1/x)$, and if $g_R(x) = a(x)b(x)$, then

$$g(x) = x^t \cdot a(1/x) \cdot b(1/x)$$

where $x^{\deg a(x)} \cdot a(1/x) \in F[x]$ and $x^{\deg b(x)} \cdot b(1/x) \in F[x]$.

Example 26.

We factor $x^{11}-1$ over $GF(3)$. The cyclotomic cosets of 3 modulo 11 are

$$C_0 = \{0\}$$

$$C_1 = \{1,3,9,5,4\}$$

$$C_2 = \{2,6,7,10,8\}.$$

Hence $x^{11} - 1$ factors as a linear and 2 irreducible quintics over $GF(3)$. We could find these quintics by working in $GF(3^5)$ and proceeding by one of the methods illustrated in example 24; however, these approaches require construction of either a Zech's log table for $GF(3^5)$, or explicit construction of this 243-element field. We may also proceed in the following somewhat ad hoc fashion. If α is an nth root of unity, then so is α^{-1}, since

$$1 = (\alpha^1\alpha^{-1})^n = \alpha^n\alpha^{-n} = \alpha^{-n}.$$

Hence if $m_\alpha(x)$ is a factor of $x^n - 1$, then so is $m_{\alpha^{-1}}(x)$. It now follows with the two observations above that the quintics we are looking for are reciprocal polynomials of each other. If one is

$$a(x) = a + bx + cx^2 + dx^3 + ex^4 + fx^5$$

then the other is

$$b(x) = f + ex + dx^2 + cx^3 + bx^4 + ax^5$$

or a scalar multiple of $b(x)$, say $\lambda b(x)$ where $\lambda = 1$ or -1. Since

$$a(x)\lambda b(x) = (x^{11}-1)/(x-1)$$

$$= x^{10} + x^9 + x^8 + \ldots + x^3 + x^2 + x + 1,$$

we get the following system of equations.

$$\lambda af = 1$$

$$\lambda(ae+bf) = 1$$

$$\lambda(ad+eb+fc) = 1$$

$$\lambda(ac+db+ec+df) = 1$$

$$\lambda(ab+cb+dc+ed+fe) = 1$$

$$\lambda(a^2+b^2+c^2+d^2+e^2+f^2) = 1.$$

We can assume without loss of generality that $a = 1$. If we suppose $\lambda = 1$, then $f = 1$ and the first 4 equations show that the only possible solutions are

e	b	d	c
0	1	0	1
1	0	1	0

Neither of these satisfy the remaining equations. Hence $\lambda = -1$ and $f = -1$. From this, we immediately deduce that e=0, b=1, d=1, c=2 is a solution and gives

$$a(x) = 2 + 2x + x^2 + 2x^3 + x^5$$

$$b(x) = 1 + 2x^2 + x^3 + 2x^4 + 2x^5$$

$$a(x)(-b(x)) = (x^{11}-1)/(x-1).$$

5.9 Another Method for Factoring x^n-1 over $GF(q)$†

In this section we present an alternate method for factoring $x^n - 1$. We require a few preliminary results. The *greatest common divisor* (gcd) of two polynomials is defined in a manner analogous to that for two integers. For polynomials $a(x), b(x) \in F[x]$, the greatest common divisor of $a(x)$ and $b(x)$ is defined to be the monic polynomial of largest degree in $F[x]$ which divides both $a(x)$ and $b(x)$. We use the notation gcd($a(x)$, $b(x)$) or simply ($a(x)$,$b(x)$).

The following result shall prove to be central to the factorization technique.

† This section may be omitted without loss of continuity.

Theorem 5.15. Let $f(x)$ be a monic polynomial of degree n over $F = GF(q)$. Let $g(x)$ be a polynomial over F with $\deg g(x) \le n-1$ and satisfying $[g(x)]^q \equiv g(x) \pmod{f(x)}$. Then

$$f(x) = \prod_{s \in F} \gcd(f(x), g(x)-s).$$

Proof.

Certainly $\gcd(f(x), g(x)-s)$ divides $f(x)$ for any $s \in F$. Now since $\gcd(a,b) = \gcd(a, b-a)$, $\gcd(g(x)-s, g(x)-t) = (g(x)-s, s-t) = 1$ for $s \neq t$. It follows that $\gcd(\gcd(f(x), g(x)-s), \gcd(f(x), g(x)-t)) = 1$ for $s \neq t$, and hence

$$\prod_{s \in F} \gcd(f(x), g(x)-s)$$

divides $f(x)$. By definition of $g(x)$, $f(x)$ divides $[g(x)]^q - g(x)$. Now note (using Lemma 2.4) that

$$y^q - y = \prod_{s \in F} (y-s)$$

over F. It follows that

$$[g(x)]^q - g(x) = \prod_{s \in F} (g(x)-s)$$

and $f(x)$ divides $\Pi_{s \in F}(g(x)-s)$. But this implies that $f(x)$ divides

$$\prod_{s \in F} \gcd(f(x), g(x)-s).$$

Since $f(x)$ and $\Pi_{s \in F} \gcd(f(x), g(x)-s)$ are both monic, we conclude

$$f(x) = \prod_{s \in F} \gcd(f(x), g(x)-s).$$

□

Example 27.

Consider q=2, n=7, $f(x) = x^7-1$ and $g(x) = x + x^2 + x^4$. Clearly $g^2(x) \equiv g(x) \pmod{f(x)}$. Using the Euclidean algorithm for polynomials (see Appendix B), it is easy to compute $\gcd(f(x), g(x)) = 1 + x + x^3$ and $\gcd(f(x), g(x)+1) = (1+x)(1+x^2+x^3)$, and to check that

$$f(x) = \gcd(f(x), g(x)) \cdot \gcd(f(x), g(x){+}1)$$
$$= (1{+}x{+}x^3)(1{+}x)(1{+}x^2{+}x^3).$$

We notice in Theorem 5.15 that if $g(x)$ has positive degree, then the factorization of $f(x)$ must be non-trivial. This follows since $\deg(f(x), g(x){-}s) < n$ if $g(x){-}s \neq 0$.

One question immediately comes to mind. How many polynomials $g(x)$ are there which satisfy $[g(x)]^q \equiv q(x) \pmod{f(x)}$?

Theorem 5.16. Let $f(x)$ be a polynomial over $F = GF(q)$ which has t distinct irreducible factors over F. Then there are exactly q^t polynomials $g(x)$ over F of degree less than n which satisfy $[g(x)]^q \equiv g(x) \pmod{f(x)}$.

Proof.

Let $f(x)$ have degree n, and let $f(x) = \Pi_{i=1}^{t} p_i^{\alpha_i}(x)$ be the complete factorization of $f(x)$ into powers of irreducible polynomials $p_i(x)$. Consider the simultaneous congruences

$$\begin{aligned} g(x) &\equiv s_1 \pmod{p_1^{\alpha_1}(x)} \\ g(x) &\equiv s_2 \pmod{p_2^{\alpha_2}(x)} \\ &\;\;\vdots \\ g(x) &\equiv s_t \pmod{p_t^{\alpha_t}(x)}, \end{aligned} \qquad (1)$$

where $s_i \in F$, $1 \le i \le t$. By the *Chinese remainder theorem* (see Appendix C), there exists a unique polynomial $g(x)$ having degree less than n which satisfies this system for any choice of s_i's. Since we can select the s_i's in q^t distinct ways, there are q^t distinct polynomials $g(x)$ satisfying the system. From (1) we get that $f(x)$ divides $\Pi_{i=1}^{t}(g(x){-}s_i)$, for any choice of the s_i, $1 \le i \le t$. Since the $p_i(x)$ are distinct irreducibles and hence coprime, each factor $g(x){-}s_k$ in this product is needed at most once in order for $f(x)$ to be a divisor. Hence $f(x)$ divides $\Pi_{s \in F}(g(x){-}s)$. But $\Pi_{s \in F}(g(x){-}s) = [g(x)]^q - g(x)$. We conclude that

$$[g(x)]^q \equiv g(x) \pmod{f(x)}. \qquad (2)$$

(Alternatively, since $g(x) \equiv s_i \pmod{p_i^{\alpha_i}(x)}$, it follows that

$$[g(x)]^q \equiv s_i^q \equiv s_i \equiv g(x) \pmod{p_i^{\alpha_i}(x)}$$

for all i, since $s_i \in F$. Then using the Chinese remainder theorem, (2) follows.)

We have established that there are at least q^t polynomials $g(x)$ with $\deg g(x) < n$ satisfying (2). The above arguments can be reversed to prove that any $g(x)$ satisfying (2) satisfies the system (1) for some choice of $s_1, s_2, ..., s_t$. Therefore, there are exactly q^t polynomials of the desired type. □

The set G of all polynomials $g(x)$ such that $g(x)$ satisfies (2) forms a subspace of $V_n(F)$. It follows from the preceding proof that this subspace has dimension t. This leads to the following result.

Theorem 5.17. Let $g_1(x), g_2(x), ..., g_t(x)$ be a basis for G. For $p_i^{\alpha_i}(x)$ and $p_j^{\alpha_j}(x)$ with $i \neq j$, there exists some integer k, $1 \leq k \leq t$, and distinct elements s, $t \in F$, such that $p_i^{\alpha_i}(x)$ divides $g_k(x)-s$ but not $g_k(x)-t$, and $p_j^{\alpha_j}(x)$ divides $g_k(x)-t$ but not $g_k(x)-s$.

Once this is established, we are then assured that application of Theorem 5.15 with $g_k(x)$, $1 \leq k \leq t$, will result in a complete factorization. This will become more clear with the example below. First, we give a proof of the result.

Proof.

Form the $t \times t$ matrix $M = [m_{ij}]$ where $g_j(x) \equiv m_{ij} \pmod{p_i^{\alpha_i}(x)}$. Theorem 5.15 guarantees that $m_{ij} \in F$. First we show that M is non-singular. Suppose there exist scalars λ_j such that $\sum_{j=1}^{t} \lambda_j m_{ij} = 0$ for each i, $1 \leq i \leq t$. Then

$$\sum_{j=1}^{t} \lambda_j m_{ij} \equiv \sum_{j=1}^{t} \lambda_j g_j(x) \pmod{p_i^{\alpha_i}(x)},$$

and so $\sum_{j=1}^{t} \lambda_j g_j(x) \equiv 0 \pmod{f(x)}$. But $\deg g_j(x) < \deg f(x)$ for each j, $1 \leq j \leq t$, and thus $\sum_{j=1}^{t} \lambda_j g_j(x) = 0$. Since the $g_j(x)$'s are linearly independent, it follows that $\lambda_j = 0$, $1 \leq j \leq t$, and thus the columns of M are linearly independent. In other words, M is non-singular. This implies that no two rows of M are identical. Thus for $i \neq j$, rows i and j differ in some column k, i.e. there is some k such that $m_{ik} \neq m_{jk}$. The result follows. □

Example 28.

Reconsider example 24 where we factor $f(x) = x^{15}-1$ over $GF(q)$ for q=2. The cyclotomic cosets are

$$C_0 = \{0\}$$

$$C_1 = \{1,2,4,8\}$$

$$C_3 = \{3,6,9,12\}$$

$$C_7 = \{7,11,13,14\}$$

$$C_5 = \{5,10\}.$$

Each coset corresponds to an irreducible factor of $f(x)$. This tells us that the subspace G contains 2^5 elements. In the special case where $f(x)$ has the form x^n-1, it is easy to find a basis for G. We use the cyclotomic cosets to form polynomials as follows.

$$g_1(x) = 1$$

$$g_2(x) = x^1 + x^2 + x^4 + x^8$$

$$g_3(x) = x^3 + x^6 + x^9 + x^{12}$$

$$g_4(x) = x^7 + x^{11} + x^{13} + x^{14}$$

$$g_5(x) = x^5 + x^{10}.$$

Now it is immediate that $g_i^2(x) \equiv g_i(x) \pmod{f(x)}$, $1 \le i \le 5$, since the powers of x with non-zero coefficients form a cyclotomic coset. It is also easy to see that the vectors associated with the $g_i(x)$'s are linearly independent, since each power of x in $g_i(x)$ is contained in none of the others. Thus we have a basis for G.

With this, we first compute

$$f(x) \;=\; \gcd(f(x), g_2(x)) \cdot \gcd(f(x), g_2(x)-1)$$

$$= (1+x+x^3+x^7)(1+x+x^2+x^4+x^8).$$

It suffices to compute $\gcd(f(x), g_2(x))$, since $\gcd(f(x), g_2(x)-1)$ can be found by dividing $f(x)$ by $(f(x), g_2(x))$. Let $a(x) = (1+x+x^3+x^7)$ and $b(x) = 1+x+x^2+x^4+x^8$. We now compute $\gcd(a(x), g_3(x))$ and $\gcd(b(x), g_3(x))$ to refine the factorization.

$$\gcd(a(x), g_3(x)) = 1 + x^3$$

$$\gcd(a(x), g_3(x)-1) = 1 + x + x^4$$

$$\gcd(b(x), g_3(x)) = 1$$

$$\gcd(b(x), g_3(x)-1) = 1 + x + x^2 + x^4 + x^8.$$

Thus

$$f(x) = (1+x^3)(1+x+x^4)(1+x+x^2+x^4+x^8).$$

We could now check $g_4(x)$ with each of these factors. But by inspection we note that $(1+x^3) = (1+x)(1+x+x^2)$ and $(1+x+x^4)$ is irreducible. Let

$$c(x) = 1 + x + x^2 + x^4 + x^8.$$

Then we need only consider $\gcd(c(x), g_4(x))$ and $\gcd(c(x), g_4(x)-1)$.

$$\gcd(c(x), g_4(x)) = 1 + x^3 + x^4$$

$$\gcd(c(x), g_4(x)-1) = 1 + x + x^2 + x^3 + x^4,$$

and we obtain

$$f(x) = (1+x)(1+x+x^2)(1+x^3+x^4)(1+x+x^4)(1+x+x^2+x^3+x^4).$$

Since we now have $f(x)$ as the product of 5 polynomials and we know that $f(x)$ has exactly 5 irreducible factors, we are sure that this is the complete factorization.

The preceding example illustrates a general method for factoring $f(x) = x^n - 1$ over $GF(q)$. A basis for the subspace G can always be found from the cyclotomic cosets of q modulo n. Using the basis elements and appropriate gcd operations, the factors of $f(x)$ can be separated.

5.10 Exercises

Cyclic subspaces.

1. Verify that $g(x)=1+x^2+x^3+x^4$ is a monic divisor of $f(x)=x^7-1$ over $F=Z_2$, and construct the ideal generated by $g(x)$ in $F[x]/(f(x))$.

2. Determine the number of cyclic subspaces in each of the following vector spaces.

   (a) $V_8(Z_2)$ (b) $V_9(Z_2)$ (c) $V_{10}(Z_2)$ (d) $V_{15}(Z_2)$

   (e) $V_{18}(Z_2)$ (f) $V_3(Z_3)$ (g) $V_4(Z_3)$

3. Show that $V_{15}(Z_2)$ contains a cyclic subspace of dimension k for each k, $0 \le k \le 15$.

4. Consider the vector space $V_{17}(Z_2)$.

   (a) Determine the number of cyclic subspaces.

   (b) Determine all values of k, $1 \le k \le 17$, for which a cyclic subspace of dimension k exists.

   (c) Determine the number of these subspaces which have dimension 12.

   (d) Give a generator polynomial for a cyclic subspace of dimension 8, if possible.

5. Determine the number of cyclic subspaces of dimension 9 in $V_{21}(Z_2)$.

6. Determine the number of cyclic subspaces of dimension 5 in $V_8(Z_3)$. Give a generating polynomial for each one.

7. Let $g(x)=a_0+a_1x+...+a_{n-1}x^{n-1}$ be a monic polynomial over F of least degree in some cyclic subspace of $V_n(F)$. Prove that $a_0 \ne 0$.

8. Let $g(x)$ and $h(x)$ be monic divisors of x^n-1 over F. Prove that if $g(x)$ divides $h(x)$, then the cyclic subspace generated by $g(x)$ contains the cyclic subspace generated by $h(x)$.

9. (a) Determine the number of cyclic subspaces in $V_6(Z_3)$.

   (b) Determine the generator polynomial and dimension of the smallest cyclic code containing the vector $\mathbf{v}=(112\,110)$ in $V_6(Z_3)$.

10. (a) Determine the number of cyclic subspaces in $V_7(Z_2)$.

    (b) Determine the generator polynomial and the dimension of the smallest cyclic code containing each of the following vectors in $V_7(Z_2)$:

    (i) $\mathbf{v}_1=(1010\,011)$ (ii) $\mathbf{v}_2=(0011\,010)$ (iii) $\mathbf{v}_3=(0101\,001)$

11. Note the prime factorization of $x^{21}-1$ over Z_2 as given in Table 3 of this chapter.

    (a) Determine the number of cyclic subspaces of $V_{21}(Z_2)$.

    (b) Determine the number of cyclic subspaces of $V_{21}(Z_2)$ of dimension 9, and the (unsimplified) generator polynomial for one of these subspaces.

    (c) Determine whether or not any of the subspaces in (b) are self-dual.

12. Determine whether or not $g(x) = 2 + 2x^2 + x^3$ is the generator polynomial for a cyclic subspace of $V_8(Z_3)$.

Generator matrices, parity-check matrices and syndromes.

13. The polynomial $g(x) = 1 + x^3 + x^6$ generates a binary cyclic (9,3)-code C. Construct a generator matrix $G = [R\ I_3]$ for C, where R is obtained from the coefficients of remainder polynomials.

14. For the generator matrix G in the problem above, construct a parity-check matrix $H = [I_6\ R^T]$. For each of the following vectors, compute the syndrome by two methods: (i) by multiplying by H, and (ii) by dividing the associated polynomial by $g(x)$:
    (a) $\mathbf{r}_1 = (101\ 100\ 100)$ (b) $\mathbf{r}_2 = (011\ 110\ 011)$ (c) $\mathbf{r}_3 = (100\ 010\ 001)$

15. Suppose a generator matrix G of a linear code C has the property that a cyclic shift of any row of G is also a codeword. Prove that C is a cyclic code.

Encoding.

16. The polynomial $g(x) = 1 + x + x^2 + x^3 + x^6$ generates a binary cyclic (15,9)-code. Encode each of the following message polynomials using the encoding scheme of §5.5:
    (a) $1 + x^2 + x^5 + x^8$ (b) $1 + x + x^2$ (c) $x^4 + x^6 + x^7 + x^8$

17. Consider the code in the problem above.

    (a) Construct a generator matrix of the form $G = [R\ I_k]$, and encode the message polynomials using vector-matrix multiplication.

    (b) Encode the message polynomials so that the message components themselves appear in the first k components of the codewords.

Decoding.

18. A (7,4)-cyclic binary code C is generated by $g(x) = 1 + x + x^3$. Decode each of the following vectors using error-trapping.
    (a) $\mathbf{r}_1 = (1101\ 011)$ (b) $\mathbf{r}_2 = (0101\ 111)$ (c) $\mathbf{r}_3 = (0011\ 111)$ (d) $\mathbf{r}_4 = (0100\ 011)$

19. A binary cyclic (15,7) code C of distance 5 is generated by the polynomial $g(x) = 1+x^4+x^6+x^7+x^8$. Decode each of the following vectors using error-trapping.
(a) $\mathbf{r}_1 = (11011\ 11011\ 10110)$ (b) $\mathbf{r}_2 = (11111\ 10101\ 00110)$
(c) $\mathbf{r}_3 = (10111\ 00000\ 01001)$ (d) $\mathbf{r}_4 = (10101\ 00101\ 10000)$

20. Consider the code C of the problem above.
(a) Construct a parity check matrix of the form $H = [I_k\ A]$ for C.
(b) Decode the received vector $\mathbf{r} = (10111\ 11010\ 11110)$.

21. A binary cyclic (15,5)-code C is generated by the polynomial $g(x) = 1+x+x^2+x^4+x^5+x^8+x^{10}$. Using the method described in example 18 of §5.6, decode each of the following vectors.
(a) $\mathbf{r}_1 = (01000\ 01110\ 11001)$ (b) $\mathbf{r}_2 = (10010\ 11110\ 11100)$
(c) $\mathbf{r}_3 = (01111\ 11101\ 01000)$ (d) $\mathbf{r}_4 = (10001\ 11011\ 00111)$

Burst errors.

22. Let C be an (n,k) code which can correct all burst errors of length t.
(a) Prove that no non-zero burst error of length $2t$ or less is a codeword.
(b) Prove that C must have at least $2t$ parity-check symbols, i.e. $n-k \geq 2t$. (Hint: establish that any $n-k+1$ columns of a parity-check matrix H for an (n,k)-code are linearly dependent, then use part (a)). This result is known as the *Rieger bound*.

23. Let C be a binary (15,10)-code generated by $g(x) = 1+x^2+x^4+x^5$.
(a) Prove that C will correct all bursts of length 2 or less.
(b) Find the distance d of C.

24. For the code given in the problem above, decode the following received vectors.
(a) $\mathbf{r}_1 = (01011\ 00000\ 00010)$ (b) $\mathbf{r}_2 = (10000\ 10110\ 10111)$
(c) $\mathbf{r}_3 = (10101\ 10000\ 10101)$

25. Let C be a binary (15,9)-code generated by $g(x) = 1+x^3+x^4+x^5+x^6$. Suppose it is known that C can correct any burst of length 3 or less. Decode each of the following vectors.
(a) $\mathbf{r}_1 = (10101\ 11010\ 11100)$ (b) $\mathbf{r}_2 = (11000\ 01110\ 10011)$
(c) $\mathbf{r}_3 = (01000\ 00010\ 11111)$

26. Prove that the code given in the problem above is not 3-error correcting.

27. Given a cyclic code C capable of correcting all burst errors of length t or less, suppose the error-trapping algorithm of §5.7 determines, for some shift $x^i r(x)$, a syndrome $s_i(x)$ with $w(s_i(x)) \le t$, but where $s_i(x)$ is not a burst of length t or less. Can any correction be made?

Factoring x^n-1.

28. Factor $f(x)=x^8-1$ over each of the following fields:

    (a) $GF(2)$ (b) $GF(3)$ (c) $GF(4)$ (d) $GF(5)$

29. Factor $f(x)=x^{10}-1$ over $GF(2)$ and $GF(3)$.

30. Factor $f(x)=x^{13}-1$ over $GF(2)$ and $GF(3)$.

General.

31. Determine whether or not the linear codes with the following generator matrices are cyclic. G_1 is binary; G_2 is over Z_3.

$$G_1 = \begin{bmatrix} 1&0&1&1&1&0&0&0 \\ 1&1&0&0&1&1&0&0 \\ 0&1&0&1&1&0&1&0 \\ 0&0&1&1&0&0&1&1 \end{bmatrix} \qquad G_2 = \begin{bmatrix} 2&0&0&2&2&2&1&0 \\ 1&0&2&2&1&2&0&0 \\ 2&1&2&1&2&1&2&1 \\ 0&1&2&1&1&2&2&2 \end{bmatrix}$$

32. A binary cyclic (15,5)-code C with distance 7 is generated by $g(x) = 1+x+x^2+x^4+x^5+x^8+x^{10}$.

    (a) Using error-trapping, determine (if possible) the error pattern associated with the vector $\mathbf{r} = (11100\ 00101\ 11011)$.

    (b) Determine the maximum number of errors that can be corrected in general using error-trapping on this code. Explain why the error in (a) was correctable (not correctable).

    (c) A generator matrix of the form $[R\ I_5]$ encodes the message k-tuple $\mathbf{m} = (10110)$ to the codeword $\mathbf{c} = (\mathbf{p},\mathbf{m})$ where the parity-check symbols are given by $\mathbf{p} = (p_0 p_1 \ldots p_9)$. Using the encoding algorithm for binary cyclic codes, determine $\mathbf{p}$.

    (d) Construct a generator matrix G for C of the form $[R\ I_5]$, where R is obtained from the coefficients of remainder polynomials.

33. (a) Prove that $g(x) = 1+x^2+x^3+x^6+x^8+x^{11}+x^{12}$ generates a binary cyclic (21,9)-code.

(b) Find the parity check polynomial $h(x)$ for the code C of part (a).

(c) Suppose it is known that C can correct all burst errors of length 6 or less. Find the burst error pattern associated with the vector $\mathbf{r} = (101\ 110\ 011\ 011\ 000\ 010\ 011)$.

34. (a) Determine the number of cyclic subspaces of dimension 3 in $V_9(Z_2)$.

(b) Give a generator polynomial $g(x)$ for a cyclic (9,3)-code C.

(c) Display a generator matrix G for C.

(d) Display a parity check matrix H for C.

(e) Determine the distance of C.

35. Let $g(x)$ be the generator for a binary cyclic (n,k)-code C.

(a) Prove that if $(x-1)$ divides $g(x)$, then all codewords have even weight.

(b) Suppose n is odd. Prove that the all 1's vector is a codeword if and only if $(x-1)$ does not divide $g(x)$. (Hint: $(x^n-1) = (x-1)(1+x+x^2+...+x^{n-1})$ over Z_2)

36. Suppose C is a binary cyclic code of odd length. Show that C contains a codeword of odd weight if and only if the all ones vector is a codeword.

37. (a) Let $x^n - 1 = g(x)h(x)$, and let $h_R(x)$ be the reciprocal polynomial of $h(x)$. Prove that a cyclic code C with generator polynomial $g(x)$ is self-orthogonal if and only if $h_R(x)$ divides $g(x)$.

(b) Give a generator polynomial for a self-orthogonal binary cyclic code of block length 15.

38. Let $x^n - 1 = g(x)\,h(x)$. Recall by Theorem 5.10 that if $g(x)$ generates an (n,k)-code C, then $h_R(x)$ generates $C^{\perp}$.

(a) Prove that if $h_R(x)$ divides $g(x)$, then $C \subseteq C^{\perp}$.

(b) Show that $g(x) = (1+x)(1+x+x^3)^2$ generates a binary cyclic (14,7) code which is self-dual and has distance $d \geq 4$. (Hint: $x^7 - 1 = (1+x)(1+x+x^3)(1+x^2+x^3)$, and note that $(1+x+x^3)^2$ divides $x^{14} - 1$ but does not divide $x^t - 1$ for $t < 14$)

39. The polynomial $g(x) = 1 + x^2 + x^4 + x^6 + x^7 + x^{10}$ is the generator polynomial for a binary cyclic (21,11)-code with distance 5.

(a) Use error-trapping to determine the error pattern associated with the vector $\mathbf{r} = (1100\ 1110\ 1100\ 1011\ 0000\ 0)$, and correct $\mathbf{r}$ to a codeword.

(b) Determine whether or not all error patterns of weight at most 2 can be corrected using error-trapping.

40. Let C be a binary cyclic (n,k)-code with generator polynomial $g(x)=(x+1)p(x)$, where $p(x)$ does not divide x^t-1 for any t, $1 \le t \le n-1$. An error pattern of the form $e(x)=x^i+x^{i+1}$ is called a *double-adjacent* error pattern.

 (a) Prove that no two double-adjacent error patterns can be in the same coset of a standard array for C.

 (b) Prove that C corrects all single errors and all double-adjacent errors.

 (c) Construct a generator polynomial for a binary cyclic (15,10)-code which corrects all single errors and all double-adjacent error patterns.

41. A *primitive* polynomial of degree m over $GF(2)$ is an irreducible polynomial of degree m which divides x^n-1 for $n=2^m-1$ and no smaller value of n. Consider the binary polynomial $g(x)=(x^3+1)p(x)$ where $p(x)$ is a primitive polynomial of degree m, $m>2$ and m even. Let n be the smallest integer such that $g(x)$ divides x^n-1, so that $g(x)$ generates a cyclic code C of length n.

 (a) Show that C is capable of correcting all single errors, double-adjacent errors and triple-adjacent errors.

 (b) Show that (a) is not necessarily true, if $p(x)$ is not primitive.

42. Let α be a primitive element in $GF(2^r)$, and let H be the parity-check matrix for a binary Hamming code of order r with column i, $0 \le i \le 2^r-2$ being the binary r-tuple corresponding to α^i. Let C be the extended code formed by adding an overall parity check to the end of this code.

 (a) Prove that C is capable of correcting all single errors and all double-adjacent errors which do not include the overall parity check position. (Note: Adding the parity check results in a code which is not cyclic.)

 (b) Can the columns of H be rearranged in such a way that C can correct all single and double-adjacent errors?

43. Determine the smallest block length n for a binary cyclic code generated by the irreducible polynomial $g(x)=1+x^4+x^5$.

44. Determine the smallest block length n for a binary cyclic code generated by the irreducible polynomial $g(x)=1+x+x^9$.

45. Let C be a binary cyclic code of length n with generator polynomial $g(x)$, where n is the smallest positive integer for which $g(x)$ divides $x^n - 1$.
(a) Show that C has minimum weight at least 3.
(b) Does (a) hold true for non-binary codes?

46. Let $g(x)$ be the generator polynomial for a binary cyclic code C. The set C_E of even weight vectors in C is also a code. Is C_E a cyclic code? If so, give a generating polynomial for C_E.

47. Let C be a $(q+1,2)$-code over $GF(q)$ with minimum distance q. If q is odd show that C cannot be cyclic. (Hint: see exercise 36, Chapter 3)

48. Let C be a Hamming code of order 2 over $GF(q)$. Show that if q is odd, then C is not equivalent to a cyclic code.

49. Given a field $GF(q)$ and an integer n such that $\gcd(n,q) = 1$, define the relation R on the integers by the rule $a\,R\,b$ if and only if for some integer j, $a \equiv bq^j \pmod{n}$. Show that this is an equivalence relation on the integers modulo n, and that the equivalence classes are the cyclotomic cosets of q mod n. (cf. exercise 51, Chapter 3)

50. Let $f(x) \in GF(q)[x]$, and let $f'(x)$ denote the derivative of $f(x)$. Prove that $f(x)$ has no repeated factors if and only if $\gcd(f(x), f'(x)) = 1$.

51. Let C be a binary (15,9)-code generated by $g(x) = 1 + x + x^2 + x^3 + x^6$. Encode each of the following message polynomials using the encoding algorithm of §5.5.

(i) $a_1(x) = 1 + x^3 + x^8$ (ii) $a_2(x) = x^3 + x^5 + x^6$

(iii) $a_3(x) = 1 + x^2 + x^5 + x^7 + x^8$.

52. Let $g(x) = 1 + x + x^3 + x^4 + x^5 + x^7 + x^8$ be a generator polynomial for a binary cyclic (15,7)-code C.

(a) Construct a parity-check matrix $H = [I_8\ R^T]$ for C where R is the matrix formed from remainder polynomials.

(b) Using the encoding algorithm of §5.5, encode

(i) $\mathbf{a}_1 = (1\ 0\ 1\ 1\ 0\ 0\ 1)$ to $\mathbf{c}_1$.

(ii) $\mathbf{a}_2 = 90\ 0\ 1\ 0\ 1\ 1\ 1)$ to $\mathbf{c}_2$.

(c) Verify that $\mathbf{c}_1$ and $\mathbf{c}_2$ are indeed two codewords in C using the H constructed in (a).

53. Let $g(x)$ be the generator polynomial for a cyclic code C over $GF(q)$ of block length n. Suppose $gcd(q,n)=1$. Prove that the all ones vector is a codeword if and only if $(x-1)$ does not divide $g(x)$. (This generalizes exercise 35 (b).)

54. Let C_1 be a binary (15,10) cyclic code generated by $g_1(x)=(x+1)(x^4+x^3+1)$ and C_2 be another binary (15,10) cyclic code generated by $h(x)=(x+1)(x^4+x+1)$.

(i) Is $(11011\ 00101\ 00000) \in C_1 \cap C_2$?

(ii) What is the generator polynomial for the code $C = C_1 \cap C_2$?

(iii) How many cyclic subspaces are contained in C_1?

55. Determine the generator polynomial and the dimension of a smallest cyclic subspace in $V_{15}(Z_2)$ which contains both $\mathbf{v}_1 = (01000\ 10101\ 00011)$ and $\mathbf{v}_2 = (10110\ 10110\ 10110)$.

56. Let C be a cyclic (n,k)-code over F generated by the polynomial $g(x)$. Prove that C will detect any burst error of length $n-k$ or less. (Such codes are called *cyclic redundancy codes* or *CRCs*.)

57. Let C be a cyclic (n,k)-code over F. Suppose each error pattern of weight less than or equal to t contains a cyclic run of at least k 0's. Let $\mathbf{r}=(r_0,r_1,\ldots,r_{n-1})$ be a received vector. For each vector $\mathbf{x}=(x_0,x_1,\ldots,x_{n-1})$ define $[\mathbf{x}]_k$ to be the k-tuple $(x_0,x_1,\ldots,x_{k-1})$. Let G be a generator matrix for C which is in standard form. Consider the following decoding algorithm: for each i, $0 \le i \le n-1$, do

1. Compute $\mathbf{c}_i = [x^i\mathbf{r}]_k G$.

2. Compute $d(\mathbf{c}_i,\mathbf{r})$. If $d(\mathbf{c}_i,\mathbf{r}) \le t$ then decode $\mathbf{r}$ to $x^{-i}\mathbf{c}_i$.

Prove that the algorithm corrects all error patterns of weight t or less.

Chapter 6

BCH CODES and BOUNDS FOR CYCLIC CODES

6.1 Introduction

A monic polynomial $g(x) \in GF(q)[x]$ is said to *split* in the extension field $GF(q^m)$ of $GF(q)$ if $g(x)$ can be factored as a product of linear polynomials in $GF(q^m)$, i.e. if we can write

$$g(x) = (x - \alpha_1)(x - \alpha_2) \cdots (x - \alpha_n)$$

where $\alpha_i \in GF(q^m)$. Here $GF(q^m)$ is called a *splitting field* of $g(x)$. In general, we refer to the splitting field of a polynomial $g(x) \in GF(q)[x]$ as the smallest field $GF(q^m)$ in which $g(x)$ splits, or in other words, the smallest field containing all roots of $g(x)$. It is not difficult to prove that the splitting field of a polynomial always exists (e.g. see [Lidl 84]). Indeed, the splitting field of $g(x)$ can be determined from the degrees of its irreducible factors over $GF(q)$ (see exercise 29). Note that $g(x)$ may be irreducible over $GF(q)$, but (always) factors as a product of distinct linear polynomials in its splitting field. For example, $g(x) = x^2 + x + 1$ is irreducible over $GF(2)$ and has no roots in $GF(2)$; but over $GF(4)$,

$$g(x) = (x + \alpha)(x + \alpha^2)$$

and has roots α and α^2, where $GF(4) = \{0,1,\alpha,\alpha^2\}$ with $\alpha^2 + \alpha + 1 = 0$.

We discuss in this chapter an important relation between the set of roots of a generator polynomial in its splitting field and the distance of the corresponding cyclic code. We will see that by constructing a generator polynomial which has a prescribed set of roots, bounds can be put on the distance of the cyclic code it generates. In this first section we present some preliminary ideas and examples to facilitate the introduction and discussion of BCH codes in §6.2. We begin with two examples illustrating the idea of the splitting field of a polynomial.

Example 1.

Consider the polynomial

$$g(x) = 1 + x^3 + x^5 + x^6 + x^8 + x^9 + x^{10}$$

over $GF(2)$. It can be verified that $g(x)$ has no roots in $GF(2)$ (check that for each $\beta \in GF(2)$, $g(\beta) \neq 0$), nor any roots in $GF(2^2)$, $GF(2^3)$ or $GF(2^4)$. Using $GF(2^5)$ as given in Appendix D, generated by the root α of $h(x) = 1 + x^2 + x^5$, it is easily verified that α and α^3 are roots of $g(x)$. It follows that the conjugates of these elements, i.e.

$$\alpha, \alpha^2, \alpha^4, \alpha^8, \alpha^{16} \text{ and } \alpha^3, \alpha^6, \alpha^{12}, \alpha^{24}, \alpha^{17}$$

are roots, that these are the only roots of $g(x)$, and that all roots of $g(x)$ are in $GF(2^5)$.

Example 2.

Consider the polynomial

$$g(x) = 2 + 2x + x^4 + 2x^5 + x^6 + x^7$$

over $GF(3)$. It can be checked that $g(x)$ has only one root in $GF(3)$, namely the element 1, and no roots aside from this in $GF(3^2)$. All roots of this polynomial lie in $GF(3^3)$. Using $GF(3^3)$ as given in Appendix D, generated by the root α of $h(x) = 1 + 2x^2 + x^3$, it is easily verified that

$$1,\ \alpha^2, \alpha^6, \alpha^{18},\ \alpha^4, \alpha^{12}, \alpha^{10}$$

are the 7 roots of $g(x)$. We shall return to this example in §6.2.

Let $g(x)$ be a generator polynomial for a cyclic (n,k)-code C over $F = GF(q)$. Then $g(x)$ is a polynomial of degree $n - k$. Let $\alpha_1, \alpha_2, \ldots, \alpha_{n-k}$ be its roots, possibly in some extension field $GF(q^m)$ of $GF(q)$; we shall assume that $g(x)$ contains no repeated factors. Let $m_i(x)$ be the minimal polynomial of α_i, with respect to $GF(q)$, $1 \le i \le n-k$. It follows, by an argument directly analogous to that immediately preceding Theorem 2.11, that $m_i(x)$ divides $g(x)$ for $1 \le i \le n-k$.

We know that a polynomial $c(x)$ is in C if and only if $g(x)$ divides $c(x)$. Hence if $c(x) \in C$, then $c(\alpha_i) = 0$, $1 \le i \le n-k$. Conversely, given any polynomial $t(x) \in F[x]$ such that $t(\alpha_i) = 0$, $1 \le i \le n-k$, then $m_i(x)$ divides $t(x)$, $1 \le i \le n-k$, implying that $g(x)$ divides $t(x)$, and hence $t(x)$ is in the code. This observation leads to the following alternate specification for the code C.

Theorem 6.1. Suppose $g(x)$ generates a cyclic (n,k)-code C over $GF(q)$, and has roots $\alpha_1, \alpha_2, \dots, \alpha_{n-k}$ in some extension field $GF(q^m)$. Then a polynomial $t(x)$ is in C if and only if $t(\alpha_i) = 0$ for all i, $1 \le i \le n-k$.

Consider the following matrix associated with code C described above.

$$H = \begin{bmatrix} \alpha_1^{\ 0} & \alpha_1^{\ 1} & \alpha_1^{\ 2} & \cdots & \alpha_1^{\ n-1} \\ \alpha_2^{\ 0} & \alpha_2^{\ 1} & \alpha_2^{\ 2} & \cdots & \alpha_2^{\ n-1} \\ \vdots & \vdots & \vdots & & \vdots \\ \alpha_{n-k}^{\ 0} & \alpha_{n-k}^{\ 1} & \alpha_{n-k}^{\ 2} & \cdots & \alpha_{n-k}^{\ n-1} \end{bmatrix}.$$

H is an $(n-k) \times n$ matrix over $GF(q^m)$. It follows that the rows of H are orthogonal to each vector in C, for if $\mathbf{c} = (c_0, c_1, \dots, c_{n-1})$ and $\mathbf{c} \in C$, then the inner product of $\mathbf{c}$ with the ith row of H gives

$$\sum_{j=0}^{n-1} \alpha_i^{\ j} c_j = c(\alpha_i) = 0.$$

Conversely, if $H\mathbf{c}^T = 0$, then $c(\alpha_i) = 0$ for all i, and hence by Theorem 6.1, $\mathbf{c} \in C$. So we now may state that $\mathbf{c} \in C$ if and only if $H\mathbf{c}^T = 0$.

Note that from H we may construct an $m(n-k) \times n$ matrix over $GF(q)$, by replacing $\alpha_i^{\ j}$ by the column vector corresponding to its m-tuple over $GF(q)$. Now as the rows of H are orthogonal to each element $\mathbf{c} \in C$, the rows of this latter matrix are also orthogonal to the codewords in C, since each element $\alpha_i^{\ j}$ can be represented as a linear combination over $GF(q)$ of m linearly independent basis elements of $GF(q^m)$. It should be noted that the rows of this latter matrix are not necessarily linearly independent over $GF(q)$.

Example 3.

$g(x) = 1 + x + x^3$ generates a binary (7,4)-code C. Recalling the argument from the introduction of §5.8, since $g(x)$ divides $x^7 - 1$ over $GF(2)$, and $2^3 \equiv 1 \pmod 7$, it follows that all roots of $g(x)$ lie in $GF(2^3)$. If we let α be a primitive element in $GF(2^3)$, with $GF(2^3)$ represented as given in Appendix D, then α, α^2 and α^4 are the roots of $g(x)$. Thus C is the set of all polynomials in $GF(2)[x]$ of degree at most 6 having α, α^2 and α^4 as roots. Of course, it is enough to say that α is a root since α^2 and α^4 are conjugate elements of α. The matrix H discussed above has first row

$$[1\ \alpha\ \alpha^2\ \alpha^3\ \alpha^4\ \alpha^5\ \alpha^6] = \begin{bmatrix} 1&0&0&1&0&1&1 \\ 0&1&0&1&1&1&0 \\ 0&0&1&0&1&1&1 \end{bmatrix},$$

associating to each power of α the corresponding binary 3-tuple. This is a parity-check matrix for C.

Example 4.

$g(x) = 1 + x^3 + x^6$ is the generator polynomial for a (9,3)-code over $GF(2)$. $g(x)$ has all of its roots in $GF(2^6)$. If α is a primitive element in $GF(2^6)$, then $\beta = \alpha^7$ is a primitive 9th root of unity. (β^i, $0 \le i \le 8$ are distinct and β^9=1.) Using $GF(2^6)$ as given in Appendix D, we see that β is one root of $g(x)$. It follows that the roots of $g(x)$ are β, β^2, β^4, β^8, β^7, β^5 and

$$H = \begin{bmatrix} (\beta)^0 & (\beta)^1 & \dots & (\beta)^8 \\ (\beta^2)^0 & (\beta^2)^1 & \dots & (\beta^2)^8 \\ \cdot & \cdot & & \cdot \\ \cdot & \cdot & & \cdot \\ \cdot & \cdot & & \cdot \\ (\beta^5)^0 & (\beta^5)^1 & \dots & (\beta^5)^8 \end{bmatrix}.$$

Example 5.

The polynomial $g(x) = 1 + x + x^2 + x^3$ generates an (8,5)-code C over $GF(3)$. $g(x)$ divides $x^8 - 1$ over $GF(3)$, and since $8 = 3^2 - 1$, all roots of $g(x)$ are in $GF(3^2)$ (recall §5.8). We construct $GF(3^2)$ using the primitive irreducible polynomial $f(x) = 2 + x + x^2$ as in example 8 of §2.5. We deduce that the roots of $g(x)$ are α^2, α^6 and α^4. Note that the minimal polynomial of α^6 is the same as that for α^2. It follows that a polynomial $c(x)$ is in C if and only if $c(\alpha^2) = 0$ and $c(\alpha^4) = 0$. The matrix H is

$$H = \begin{bmatrix} (\alpha^2)^0 & (\alpha^2)^1 & \cdots & (\alpha^2)^7 \\ (\alpha^4)^0 & (\alpha^4)^1 & \cdots & (\alpha^4)^7 \\ (\alpha^6)^0 & (\alpha^6)^1 & \cdots & (\alpha^6)^7 \end{bmatrix}.$$

Recalling Theorem 6.1, the last row of H is redundant (since if α^2 satisfies the polynomial, then so does α^6), and we can omit it in the characterization of C. Excluding the redundant row, and replacing the elements in the remaining rows of H by their corresponding column 2-tuples over $GF(3)$, yields the parity-check matrix

$$H' = \begin{bmatrix} 1 & 1 & 2 & 2 & 1 & 1 & 2 & 2 \\ 0 & 2 & 0 & 1 & 0 & 2 & 0 & 1 \\ 1 & 2 & 1 & 2 & 1 & 2 & 1 & 2 \\ 0 & 0 & 0 & 0 & 0 & 0 & 0 & 0 \end{bmatrix}.$$

From H' we see that C has distance 2.

In the next section, we apply the ideas presented here to construct a very important class of cyclic codes, the BCH codes. As implied earlier, by employing generator polynomials which contain a prescribed set of roots we obtain codes with good error-correcting capabilities. Perhaps more important, efficient and easily implemented decoding schemes are known for this class of codes.

6.2 BCH Codes and the BCH Bound

In Chapter 5 we discussed the construction of cyclic codes, but did not specify how to construct a cyclic code having a given distance. As we have seen, distance is a critical aspect of any code. With the ideas of the previous section, we are now ready to address this issue.

A very important class of cyclic codes was discovered in 1960 by R.C. Bose and D. Ray-Chaudhuri, and independently in 1959 by A. Hocquenghem. These codes are now referred to as *BCH codes* (Bose-Chaudhuri-Hocquenghem codes). A formal definition follows. By $lcm\{a(x), b(x), ...\}$ we mean the *least common multiple* of the polynomials listed. Let $m_i(x)$ denote the minimal polynomial of β^i, an element in some extension field $GF(q^m)$ of $GF(q)$.

Definition. A *BCH code* over $F = GF(q)$ of block length n and *designed distance* δ is a cyclic code generated by a polynomial $g(x) = lcm\{m_i(x): a \le i \le a+\delta-2\} \in F[x]$ whose root set contains $\delta-1$ distinct elements $\beta^a, \beta^{a+1}, ..., \beta^{a+\delta-2}$, where β is a primitive n^{th} root of unity and a is some integer.

If $n = q^m-1$ for some positive integer m, then the code is said to be *primitive*, and if $a = 1$ the code is sometimes called *narrow-sense*. Note that since β is an nth root of unity, β^j is also an nth root of unity for all j, and hence it is clear that $(x-\beta^j)$ divides (x^n-1), and that $g(x)$ divides (x^n-1), as required for the generator polynomial of a cyclic code of block length n.

In the construction of a BCH code of block length n, to find a primitive n^{th} root of unity, we recall the discussion of §5.8. One determines the extension field $GF(q^m)$ of $GF(q)$ defined by the smallest positive integer m such that n divides q^m-1. If $q^m-1 = kn$, and α is a primitive element of $GF(q^m)$, then $\beta = \alpha^k$ is a primitive n^{th} root of unity.

Example 6.

In example 3, $g(x) = 1 + x + x^3$ generates a (7,4)-BCH code with designed distance 3. Among the roots of $g(x)$ are $\beta = \alpha$ and β^2, where β is a 7th root of unity in $GF(2^3)$.

Example 7.

Consider the polynomial $g(x) = 2 + 2x + x^4 + 2x^5 + x^6 + x^7$ over $GF(3)$ of example 2. $g(x)$ divides $x^{13} - 1$, and generates a (13,6)-code over $GF(3)$. Using the finite field $GF(3^3)$ given in Appendix D, $\beta = \alpha^2$ is a primitive 13th root of unity and it is easily verified that $\beta^0, \beta^1, \beta^2, \beta^3$ are among the roots of $g(x)$. $g(x) = lcm\{m_i(x): 0 \le i \le 3\}$ where $m_i(x)$ is the minimal polynomial of β^i, since $m_0(x) = x - 1$, $m_1(x) = m_3(x) = 2 + 2x + 2x^2 + x^3$, and $m_2(x) = 2 + 2x + x^3$. Hence $g(x)$ generates a (13,6)-BCH code with designed distance 5.

The following result explains the use of the term *designed distance.*

Theorem 6.2 *(The BCH bound).* Let C be a BCH code over $GF(q)$ with designed distance δ. Then C has distance at least δ.

Proof.

Suppose $g(x)$ generates C and C has block length n. Let α be a primitive nth root of unity in $GF(q^m)$, and let $\alpha^a, \alpha^{a+1}, \ldots, \alpha^{a+\delta-2}$ be $\delta-1$ distinct elements which are among the roots of $g(x)$. Form the matrix

$$H = \begin{bmatrix} (\alpha^a)^0 & (\alpha^a)^1 & (\alpha^a)^2 & \cdots & (\alpha^a)^{n-1} \\ (\alpha^{a+1})^0 & (\alpha^{a+1})^1 & (\alpha^{a+1})^2 & \cdots & (\alpha^{a+1})^{n-1} \\ \vdots & \vdots & \vdots & & \vdots \\ (\alpha^{a+\delta-2})^0 & (\alpha^{a+\delta-2})^1 & (\alpha^{a+\delta-2})^2 & \cdots & (\alpha^{a+\delta-2})^{n-1} \end{bmatrix}.$$

Suppose we can prove that no $\delta-1$ columns of H are linearly dependent over $GF(q^m)$. Then, replacing elements of $GF(q^m)$ in H by their corresponding column m-tuples over $GF(q)$, it follows that no set of $\delta-1$ columns in this derived matrix is linearly dependent over $GF(q^m)$, or over $GF(q)$ for that matter. Furthermore, then no $\delta-1$ columns of a maximum set of linearly independent rows of this derived matrix are linearly dependent. Now by Theorem 6.1 and the discussion following it, the rows of this derived matrix are vectors in $C^{\perp}$, and a maximum set of linearly independent rows of this derived matrix yields a parity check matrix H' for C. Finally, since no $\delta-1$ columns of H' are linearly dependent over

$GF(q)$, it follows by Theorem 3.5 that C has distance at least δ, which is what we wish to prove.

To achieve this end, consider any $\delta-1$ columns of H. They form a $(\delta-1) \times (\delta-1)$ matrix over $GF(q^m)$. We prove that the determinant of this matrix is non-zero, implying the columns are linearly independent. The determinant of an arbitrary set of $\delta-1$ columns of H is

$$D = \begin{vmatrix} (\alpha^a)^{i_1} & (\alpha^a)^{i_2} & \cdots & (\alpha^a)^{i_t} \\ (\alpha^{a+1})^{i_1} & (\alpha^{a+1})^{i_2} & \cdots & (\alpha^{a+1})^{i_t} \\ \vdots & \vdots & & \vdots \\ (\alpha^{a+\delta-2})^{i_1} & (\alpha^{a+\delta-2})^{i_2} & \cdots & (\alpha^{a+\delta-2})^{i_t} \end{vmatrix}$$

where $t = \delta-1$. Pulling a common factor from each column gives

$$D = \prod_{j=1}^{t} (\alpha^a)^{i_j} \begin{vmatrix} 1 & 1 & \cdots & 1 \\ \alpha^{i_1} & \alpha^{i_2} & \cdots & \alpha^{i_t} \\ (\alpha^2)^{i_1} & (\alpha^2)^{i_2} & \cdots & (\alpha^2)^{i_t} \\ \vdots & \vdots & & \vdots \\ (\alpha^{\delta-2})^{i_1} & (\alpha^{\delta-2})^{i_2} & \cdots & (\alpha^{\delta-2})^{i_t} \end{vmatrix}.$$

This is a *Vandermonde determinant*, which can be evaluated as follows. Begin by manipulating the determinant to get a 0 in every row of the first column except row 1. Multiply row h by α^{i_1} and subtract this from row $h+1$, $0 \le h \le \delta-2$. Having done this, expand the determinant along the first column. Since column 1 is all 0 except for a single 1 in the first row, the result is a $(\delta-2) \times (\delta-2)$ determinant with a common factor of $\alpha^{i_j} - \alpha^{i_1}$ in column $j-1$. This gives

$$D = \{ \prod_{j=1}^{t} (\alpha^a)^{i_j} \} \prod_{j=2}^{t} (\alpha^{i_j} - \alpha^{i_1}) \begin{vmatrix} 1 & 1 & \cdots & 1 \\ \alpha^{i_2} & \alpha^{i_3} & \cdots & \alpha^{i_t} \\ (\alpha^2)^{i_2} & (\alpha^2)^{i_3} & \cdots & (\alpha^2)^{i_t} \\ \vdots & \vdots & & \vdots \\ (\alpha^{\delta-3})^{i_2} & (\alpha^{\delta-3})^{i_3} & \cdots & (\alpha^{\delta-3})^{i_t} \end{vmatrix}$$

The resulting determinant is also a Vandermonde determinant of smaller size. By induction we get

$$D = \prod_{j=1}^{t} (\alpha^{a})^{i_j} \prod_{j>l} (\alpha^{i_j} - \alpha^{i_l}).$$

Since α is a primitive nth root of unity $\alpha^{i_j} - \alpha^{i_l} \neq 0$ for $j \neq l$, $1 \le l \le n$, $1 \le j \le n$, and hence $D \neq 0$. This completes the proof. (Compare to Chapter 3, example 13.) □

This theorem gives us a method for constructing cyclic codes with a lower bound on the distance in the code. (Note that the distance of the code may exceed the designed distance.)

Example 8.

Suppose we wish to construct a BCH code C with block length 15 and designed distance 7 over $GF(2)$. First, we require a primitive 15th root of unity α. We can take α to be a primitive element in $GF(2^4)$. The cyclotomic cosets of 2 modulo 15 are

$C_0 = \{0\}$

$C_1 = \{1,2,4,8\}$

$C_3 = \{3,6,9,12\}$

$C_5 = \{5,10\}$

$C_7 = \{7,11,13,14\}$.

From this we see that $m_1(x)$, $m_3(x)$ and $m_7(x)$ are irreducible polynomials of degree 4 over $GF(2)$, and $m_5(x)$ is an irreducible quadratic. If we take

$$g(x) = m_1(x)\, m_3(x)\, m_5(x)$$

then $g(x)$ divides $x^{15} - 1$ and so generates a (15,5)-cyclic code. The set of roots of $g(x)$ is

$$R = \{\alpha^i : i \in \{1, 2, 4, 8, 3, 6, 9, 12, 5, 10\}\}$$

which contains the consecutive powers α, α^2, α^3, α^4, α^5, α^6 and so C is a BCH code with designed distance 7. If we take $GF(2^4)$ to be generated by a root α of $f(x) = 1 + x + x^4$ then

$$m_1(x) = 1 + x + x^4$$

$$m_3(x) = 1 + x + x^2 + x^3 + x^4$$

$$m_4(x) = 1 + x + x^2$$

and

$$g(x) = 1 + x + x^2 + x^4 + x^5 + x^8 + x^{10}.$$

Since $g(x)$ is a codeword of weight 7, the actual distance of the code is 7.

We could also have used $g(x) = m_3(x)\, m_5(x)\, m_7(x)$, as its root set contains the consecutive powers α^9, α^{10}, α^{11}, α^{12}, α^{13}, α^{14} and so the designed distance would still be 7. In this case $g(x) = 1 + x^2 + x^5 + x^6 + x^8 + x^9 + x^{10}$, and again the true distance is 7.

Example 9.

Consider BCH codes with block length 21 over $GF(2)$. To construct such a code, we require a 21st root of unity. Use $GF(2^6)$ as given in Appendix D. Since α is a primitive element in this field, $\beta = \alpha^3$ is a primitive 21st root of unity. Now the cyclotomic cosets of 2 modulo 21 are

$$C_0 = \{0\}$$

$$C_1 = \{1,2,4,8,16,11\}$$

$$C_3 = \{3,6,12\}$$

$$C_5 = \{5,10,20,19,17,13\}$$

$$C_7 = \{7,14\}$$

$$C_9 = \{9,18,15\}.$$

Let $m_i(x)$ be the minimal polynomial of β^i. If we take $g_1(x) = m_1(x)m_3(x)$, then the root set of $g_1(x)$ contains β, β^2, β^3, β^4 and the code has designed distance 5. If we take $g_2(x) = m_1(x)m_3(x)m_5(x)$, then $g_2(x)$ generates a BCH code of designed distance 7.

Example 10.

Consider BCH codes with block length 13 over $GF(3)$. We require a primitive 13th root of unity. Using $GF(3^3)$ as given in Appendix D, $\beta = \alpha^2$ is a primitive 13th root of unity. The cyclotomic cosets of 3 modulo 13 are

$$C_0 = \{0\}$$

$$C_1 = \{1,3,9\}$$

$$C_2 = \{2,6,5\}$$

$$C_4 = \{4,12,10\}$$

$$C_7 = \{7,8,11\}.$$

If $m_i(x)$ is the minimal polynomial of β, then $g_1(x) = m_0(x)\, m_1(x)\, m_2(x)$ generates a (13,6)-BCH code over $GF(3)$ with designed distance 5, as the root set of $g_1(x)$ contains $\beta^0, \beta^1, \beta^2, \beta^3$. The polynomial $g_2(x) = m_1(x)\, m_2(x)\, m_4(x)$ generates a (13,4)-BCH code over $GF(3)$ with designed distance 7.

The code generated by $g_1(x)$ in example 10 actually has distance 6 (exceeding the designed distance 5). The reader might like to prove this. (A proof is given in example 16 of §6.4.)

The next section gives a more general method for determining the distance of a cyclic code. The result given there will imply the BCH bound of Theorem 6.2.

6.3 Bounds for Cyclic Codes†

Let C be a cyclic code of length n generated by the polynomial $g(x)$ over $GF(q)$. We will be interested in the set of roots of $g(x)$; hence as in §5.8, we will assume that $\gcd(n,q) = 1$. (Otherwise, if $gcd(n,q) = p^s$, each root will have multiplicity p^s.) This simplifies the discussion without loss of generality.

Let $F = GF(q^m)$ be a field containing the roots of $x^n - 1$ and define R to be the root set of $g(x)$ in F, i.e.

† This section may be omitted without loss of continuity.

$$R = \{\, r \in F:\ g(r) = 0 \,\}.$$

Then

$$g(x) = lcm\{\, m_r(x): r \in R \,\}.$$

In this section, we consider a useful result, distinct from Theorem 6.2, that establishes a lower bound on the distance of C. First, we require the following definition, and provide an example.

Definition. Let S be a particular subset of F. Denote by I_S the family of subsets of F defined inductively by the following rules.

(1) $\varnothing \in I_S$, where $\varnothing$ denotes the empty set.

(2) If $A \in I_S$, $A \subseteq S$ and $b \notin S$, then $A \cup \{b\} \in I_S$.

(3) If $A \in I_S$ and $c \in F$, $c \neq 0$, then $cA = \{ca: a \in A\} \in I_S$.

Each member of I_S is said to be *independent with respect to S*.

Example 11.

Consider $F = GF(2^4)$ with α a primitive element. Let $S = \{\alpha^1, \alpha^2, \alpha^5, \alpha^7\}$. Then $F \backslash S = \{0, \alpha^3, \alpha^4, \alpha^6, \alpha^8, \alpha^9, \alpha^{10}, \alpha^{11}, \alpha^{12}, \alpha^{13}, \alpha^{14}\}$ and

(i) $\varnothing \in I_S$

(ii) $A_1 = \varnothing \cup \{\alpha^3\} = \{\alpha^3\} \in I_S$

(iii) $A_2 = \alpha^{-1}A_1 \cup \{\alpha^3\} = \{\alpha^2, \alpha^3\} \in I_S$ since $\alpha^{-1}A_1 \in I_S$ and $\alpha^{-1}A_1 \subseteq S$.

(iv) $A_3 = \alpha^{-1}A_2 \cup \{\alpha^{11}\} = \{\alpha^1, \alpha^2, \alpha^{11}\} \in I_S$ since $\alpha^{-1}A_2 \in I_S$ and $\alpha^{-1}A_2 \subseteq S$.

Theorem 6.3. Let $h(x)$ be a polynomial over $GF(q)$ and let S be the set of roots of $h(x)$ in $F = GF(q^m)$. (It is not required that *all* roots of $h(x)$ be in F.) Then the number of non-zero terms in $h(x)$, denoted $\text{wt}(h(x))$, satisfies

$$\text{wt}(h(x)) \geq |A|$$

for every subset A of F which is independent with respect to S (i.e. for all $A \in I_S$).

Proof.

Suppose $h(x)$ has k non-zero terms, and

$$h(x) = \lambda_1 x^{i_1} + \lambda_2 x^{i_2} + \cdots + \lambda_k x^{i_k}$$

where $\lambda_j \in GF(q)$, $\lambda_j \neq 0$, $1 \leq j \leq k$. For any subset A of F define a set of vectors $V(A)$ in $V_k(F)$ by

$$V(A) = \{\, (a^{i_1}, a^{i_2}, \ldots, a^{i_k}):\ a \in A \,\}.$$

We prove inductively that if $A \in I_S$ then the set of vectors $V(A)$ is linearly independent in $V_k(F)$ and $|V(A)| = |A|$. Then since $\dim(V_k(F)) = k$, we have

$$|A| = |V(A)| \leq k = \mathrm{wt}(h(x)),$$

establishing the result. To begin, if $A = \varnothing$, then the result is true. Now suppose the result is true for A. If $A \in I_S$, $A \subseteq S$, and $b \notin S$, then $A \cup \{b\} \in I_S$. Since $b \notin S$, $h(b) \neq 0$ and it follows that the vector $\mathbf{b} = (b^{i_1}, b^{i_2}, \ldots, b^{i_k})$ is distinct from any vector in $V(A)$, since $h(a) = 0$ for all $a \in A$. Suppose $\mathbf{b}$ is an F-linear combination of the vectors in $V(A)$. We know that $\lambda = (\lambda_1, \lambda_2, \ldots, \lambda_k)$ has the property that $\lambda \cdot \mathbf{a} = 0$ for all $\mathbf{a} \in V(A)$. If $\mathbf{b}$ is a linear combination of the vectors in $V(A)$, then $\lambda \cdot \mathbf{b} = 0$. But this implies that $h(b) = 0$ and $b \in S$ which is not so. Therefore $V(A \cup \{b\})$ is an independent set of vectors in $V_k(F)$. Finally, if $A \in I_S$ and $c \in F$, $c \neq 0$, then $cA \in I_S$ and

$$V(cA) = \{\, (\,(ca)^{i_1}, (ca)^{i_2}, \ldots, (ca)^{i_k}\,): a \in A \,\}.$$

Consider the matrix

$$D = \begin{bmatrix} c^{i_1} & & 0 \\ & c^{i_2} & \\ & & \\ 0 & & c^{i_k} \end{bmatrix}.$$

Since $\det D \neq 0$, D represents a linear transformation from $V_k(F)$ to $V_k(F)$. D carries the independent vectors $V(A)$ into the set $V(cA)$ which implies that $V(cA)$ is a linearly independent set with $|V(A)| = |V(cA)|$. This completes the proof. □

We can apply this result to determine a lower bound, better than that provided by Theorem 6.2, on the distance of a cyclic code C generated by a polynomial $g(x)$. Since C is a linear code, its distance is equal to the minimum weight among its non-zero codewords. If we can show that every non-zero codeword (polynomial) has at least d non-zero terms, then we will know that C has distance at least d.

Example 12.

Consider binary cyclic codes of block length 17. If $g(x)$ is a generator, then $g(x)$ divides $x^{17}-1$ and the roots of $g(x)$ lie in some extension field $GF(2^m)$. Consider the cyclotomic cosets of 2 modulo 17. They are

$$C_0 = \{0\}$$

$$C_1 = \{1, 2, 4, 8, 16, 15, 13, 9\}$$

$$C_3 = \{3, 6, 12, 7, 14, 11, 5, 10\}.$$

Suppose $g(x) = m_1(x)$ where $m_1(x)$ is the minimal polynomial of α, a primitive 17th root of unity. The BCH bound tells us only that $g(x)$ has distance $d \geq 3$ (since only two roots are consecutive powers of α). To derive a better bound on the distance, we proceed as follows. Define the sequence $(A_i : i \geq 0)$ of sets independent with respect to $S = \{\alpha^i : i \in C_1\}$ by $A_0 = \varnothing$, $A_{i+1} = c_i A_i \cup \{b_i\}$. Through judicious selection of b_i and c_i, we have

$A_0 = \varnothing$		$b_0 = \alpha^3$
$A_1 = \{\alpha^3\}$	$c_1 = \alpha^1$	$b_1 = \alpha^{14}$
$A_2 = \{\alpha^4, \alpha^{14}\}$	$c_2 = \alpha^4$	$b_2 = \alpha^7$
$A_3 = \{\alpha^8, \alpha^1, \alpha^7\}$	$c_3 = \alpha^1$	$b_3 = \alpha^6$
$A_4 = \{\alpha^9, \alpha^2, \alpha^8, \alpha^6\}$	$c_4 = \alpha^7$	$b_4 = \alpha^3$
$A_5 = \{\alpha^{16}, \alpha^9, \alpha^{15}, \alpha^{13}, \alpha^3\}.$		

Since $|A_5| = 5$, wt$(g(x)) \geq 5$ (by Theorem 6.3). Now every code polynomial $c(x)$ contains S in its set of roots. Hence using the above sequence A_i, it can be shown that any polynomial $c(x)$ not containing any of α^3, α^{14}, α^7, α^6 among its roots also satisfies wt$(c(x)) \geq 5$. If $c(x)$ contains a root from the set C_3, then $c(x)$ must be divisible by $m_3(x)$, and hence $c(x) = m_1(x)\, m_3(x)$ or $c(x) = (x-1)\, m_1(x)\, m_3(x) \equiv 0$. If $c(x) = m_1(x)\, m_3(x)$, then wt$(c(x)) = 17$. Hence in all cases, wt$(c(x)) \geq 5$ for $c(x) \neq 0$, implying $d \geq 5$ (by Theorem 3.1).

Theorem 6.3 allows us to prove the following result.

> **Theorem 6.4.** If $h(x)$ has $\gamma\alpha^i$, $0 \le i \le k-1$ among its roots for some $\gamma, \alpha \in F$, then either
>
> (i) $\mathrm{wt}(h(x)) \ge k+1$, or
>
> (ii) $h(\gamma\alpha^i) = 0$ for all $i \ge 0$.

Proof.

Let $S = \{\rho \in F: h(\rho) = 0\}$. Suppose (ii) does not hold and let $i = t$ be the smallest i such that $\gamma\alpha^t \notin S$. (This may necessitate increasing the value of k.) Define the sequence $(A_i: i \ge 0)$ of subsets of F, where $A_0 = \varnothing$ and for $i \ge 0$, $A_{i+1} = \alpha^{-1}A_i \cup \{\gamma\alpha^t\}$. Then $A_0, A_1, ..., A_{k+1}$ are all independent with respect to S. Since $|A_{k+1}| = k+1$, the result follows from Theorem 6.3. □

Example 13.

Consider the binary cyclic (35,28)-code C generated by $g(x) = m_{\beta^5}(x) \cdot m_{\beta^7}(x)$ where $\beta \in GF(2^{12})$ is a primitive 35th root of unity. The root set R of $g(x)$ is

$$R = \{\beta^5, \beta^{10}, \beta^{20}, \beta^7, \beta^{14}, \beta^{28}, \beta^{21}\}.$$

By the BCH bound C has distance $d \ge 3$. Let $c(x)$ be any codeword and let S be its set of roots. If we take $\gamma = \beta^7$ and $\alpha = \beta^7$ and apply Theorem 6.4, then either $\mathrm{wt}(c(x)) \ge 5$ since $\{\beta^7, \beta^{14}, \beta^{28}, \beta^{21}\} \subseteq S$, or $\alpha^i \in S$ for all $i \ge 1$. This last possibility implies that $\beta^0 = 1 \in S$, and if 1 is a root of $c(x)$, then $\mathrm{wt}(c(x))$ is even. In either case, $\mathrm{wt}(c(x)) \ge 4$, so the distance of the code is at least 4.

Example 14.

Consider $F = GF(2^4)$ with primitive element α. Let $g(x) = m_{\alpha^1}(x)\, m_{\alpha^3}(x)$ be the generator polynomial for a (15,7)-code C. It can be established that C is a BCH code with designed distance 5 by Theorem 6.2. We could also establish this bound on the distance as follows. The roots of $g(x)$ form the set $S = \{\alpha^1, \alpha^2, \alpha^3, \alpha^4, \alpha^6, \alpha^8, \alpha^9, \alpha^{12}\}$. We construct a sequence of sets $(A_i: i \ge 0)$ which are independent with respect to S. For each i, $0 \le i \le 4$, $c_iA_i \subseteq S$, $b_i \notin S$ and $A_{i+1} = c_iA_i \cup \{b_i\}$. We use $c_i = \alpha^{-1}$ and $b_i = \alpha^5$.

$$A_0 = \varnothing$$
$$A_1 = \{\alpha^5\}$$
$$A_2 = \{\alpha^4, \alpha^5\}$$
$$A_3 = \{\alpha^3, \alpha^4, \alpha^5\}$$
$$A_4 = \{\alpha^2, \alpha^3, \alpha^4, \alpha^5\}$$
$$A_5 = \{\alpha^1, \alpha^2, \alpha^3, \alpha^4, \alpha^5\}$$

Since $|A_5| = 5$, by Theorem 6.3 $\text{wt}(g(x)) \geq 5$. Now since we used $b_i = \alpha^5$, and we require $b_i \notin S$, for any code polynomial $c(x)$ with $c(\alpha^5) \neq 0$ the same sequence A_i yields $\text{wt}(c(x)) \geq 5$. If α^5 is a root of $c(x)$, and α^7 is not (note α^6 is a root then – see example 8 for the cyclotomic cosets of 2 mod 15), then a similar sequence A_i using $b_i = \alpha^7$ yields $\text{wt}(c(x)) \geq 5$. If α^5 and α^7 are both roots of $c(x)$, then as in example 12, $\text{wt}(c(x)) = 0$ or 15. We conclude that $d \geq 5$.

Theorem 6.4 and example 14 suggest an alternate proof of Theorem 6.2 (exercise 2).

6.4 Decoding BCH Codes

There are several techniques for decoding BCH codes. The method described here is chosen because of its simplicity and mathematical elegance. We first consider the case of narrow-sense codes (i.e. $a = 1$).

Let $g(x)$ be the generator polynomial for a BCH code over $F = GF(q)$ of designed distance δ and length n. Let $\alpha^1, \alpha^2, \ldots, \alpha^{\delta-1}$ be $\delta-1$ consecutive roots of $g(x)$, where α is a primitive nth root of unity, and let $m_i(x)$ be the minimal polynomial of α^i, $1 \leq i \leq \delta-1$. As we have seen

$$g(x) = lcm\{\, m_i(x) : 1 \leq i \leq \delta-1 \,\}.$$

Suppose $c(x)$ is a code polynomial and $e(x)$ is an error polynomial with $\text{wt}(e(x)) = l \leq t$, where $t = \lfloor (\delta-1)/2 \rfloor$. Consider $r(x) = c(x) + e(x)$. Given $r(x)$, we wish to deduce $e(x)$.

First we compute the syndrome $s(x)$ of $r(x)$ by dividing $r(x)$ by $g(x)$, giving

$$r(x) = h(x)g(x) + s(x).$$

Define $S_i = s(\alpha^{i+1})$, $0 \leq i \leq \delta-2$ and compute these quantities. Clearly, $s(\alpha^{i+1}) = r(\alpha^{i+1})$. Furthermore, $r(x) = c(x) + e(x)$ implies $r(\alpha^{i+1}) = e(\alpha^{i+1})$, $0 \leq i \leq \delta-2$ and

$$S_i = r(\alpha^{i+1}) = e(\alpha^{i+1}).$$

Note that S_i can be computed by dividing $r(x)$ by $m_{i+1}(x)$ and then evaluating the remainder at α^{i+1}.

Now let

$$e(x) = \sum_{j=0}^{l-1} \lambda_{a_j} x^{a_j}$$

where $0 \le a_j \le n-1$ and $\lambda_{a_j} \ne 0$, and let $u_j = \alpha^{a_j}$, $0 \le j \le l-1$. The u_j are called *error locators* and the a_j are *error location numbers*, specifying the coordinate positions of the errors (i.e. the non-zero terms of $e(x)$). As a first goal, we set out to determine these error locators u_j as follows. Letting $y_j = \lambda_{a_j}$, $0 \le j \le l-1$ we have

$$S_i = \sum_{j=0}^{l-1} y_j u_j^{i+1}.$$

We define the *error locator polynomial* $\sigma(z)$ to be

$$\sigma(z) = \prod_{j=0}^{l-1} (1 - u_j z).$$

Note that to distinguish this polynomial from others used previously, we make a point of switching to the indeterminate z. The reciprocals of the roots of $\sigma(z)$ are then the error locators u_j, $0 \le j \le l-1$. We next define the *error evaluator polynomial* $w(z)$ to be

$$w(z) = \sum_{i=0}^{l-1} y_i u_i \prod_{\substack{j=0 \\ j \ne i}}^{l-1} (1 - u_j z).$$

Now

$$\begin{aligned}
\frac{w(z)}{\sigma(z)} &= \sum_{i=0}^{l-1} \frac{y_i u_i}{(1 - u_i z)} = \sum_{i=0}^{l-1} y_i u_i \sum_{j=0}^{\infty} (u_i z)^j \\
&\equiv \sum_{i=0}^{l-1} y_i u_i \sum_{j=0}^{\delta-2} (u_i z)^j \pmod{z^{\delta-1}} \\
&= \sum_{j=0}^{\delta-2} \sum_{i=0}^{l-1} y_i u_i^{j+1} z^j
\end{aligned}$$

$$= \sum_{j=0}^{\delta-2} S_j z^j.$$

Defining the *syndrome polynomial* $S(z)$ to be

$$S(z) = \sum_{j=0}^{\delta-2} S_j z^j,$$

a polynomial in z with coefficients from some extension field of $GF(q)$, we obtain the *key equation*

$$w(z) \equiv \sigma(z)S(z) \pmod{z^{\delta-1}} \tag{1}$$

where $\deg w(z) = l-1$, $\deg\sigma(z) = l$ and $\deg S(z) \le \delta-2$.

Suppose now that given $S(z)$ we can find non-zero polynomials $w'(z)$ and $\sigma'(z)$ such that $\deg w'(z) \le l-1$, $\deg\sigma'(z) \le l$, and

$$w'(z) \equiv \sigma'(z)S(z) \pmod{z^{\delta-1}}. \tag{2}$$

Multiplying (1) by $\sigma'(z)$ and (2) by $\sigma(z)$, we have

$$w(z)\sigma'(z) \equiv w'(z)\sigma(z) \pmod{z^{\delta-1}}.$$

But $\deg(w(z)\sigma'(z)) \le 2l-1 \le \delta-2$, and similarly $\deg(w'(z)\sigma(z)) \le \delta-2$. Hence

$$w(z)\sigma'(z) = w'(z)\sigma(z). \tag{3}$$

From the definitions of $w(z)$ and $\sigma(z)$, it follows easily that $\gcd(w(z), \sigma(z)) = 1$ (exercise 3). Therefore (3) implies that $w(z)$ divides $w'(z)$ and $\sigma(z)$ divides $\sigma'(z)$. Since by hypothesis $w'(z)$ and $\sigma'(z)$ are non-zero and $\deg w'(z) \le \deg w(z)$, $\deg\sigma'(z) \le \deg\sigma(z)$, we have

$$w'(z) = \lambda_1 w(z) \quad \text{and} \quad \sigma'(z) = \lambda_2\sigma(z)$$

for some $\lambda_1, \lambda_2 \in F$. Regardless, the roots of $\sigma'(z)$ are the same as the roots of $\sigma(z)$, and hence the roots of $\sigma'(z)$ determine the error vector.

What we need now is an algorithm to provide $w'(z)$ and $\sigma'(z)$ satisfying (2), given $S(z)$. The algorithm we will use is the *extended Euclidean algorithm* for polynomials (see Appendix B). Given polynomials $a(z)$ and $b(z)$ in $F[z]$, their gcd $G(z)$ can be computed by the ordinary Euclidean algorithm. The extended Euclidean algorithm provides us not only with $G(z)$, but also with polynomials $s(z)$ and $t(z)$ such that

$$s(z)\,a(z) + t(z)\,b(z) = G(z).$$

The algorithm may be run as follows. Let

$$s_0(z) = 1, \quad t_0(z) = 0, \quad r_0(z) = a(z),$$

$$s_1(z) = 0, \quad t_1(z) = 1, \quad r_1(z) = b(z).$$

Then for $i \geq 2$,

$$s_i(z) = s_{i-2}(z) - q_i(z) \cdot s_{i-1}(z), \qquad t_i(z) = t_{i-2}(z) - q_i(z) \cdot t_{i-1}(z)$$

where $q_i(z)$ is determined by applying the division algorithm to $r_{i-2}(z)$ and $r_{i-1}(z)$,

$$r_{i-2}(z) = q_i(z) \cdot r_{i-1}(z) + r_i(z)$$

with $r_i(z) = 0$ or $\deg r_i(z) < \deg r_{i-1}(z)$. It can be proven that if $\deg a(z) \geq \deg b(z)$, then for $i \geq 1$,

$$\deg t_i(z) \leq \deg a(z) - \deg r_{i-1}(z), \tag{4}$$

and that for $i \geq 0$

$$s_i(z)\,a(z) + t_i(z)\,b(z) = r_i(z) \tag{5}$$

(exercise 27). We apply this method with $a(z) = z^{\delta-1}$ and $b(z) = S(z)$, continuing iterations until $\deg r_{i-1} \geq (\delta-1)/2$ and $\deg r_i < (\delta-1)/2$. Such a remainder $r_i(z)$ exists since the degree of the polynomials in the Euclidean remainder sequence decreases monotonically, the sequence terminating with the last non-zero remainder being $G(z) = \gcd(z^{\delta-1}, S(z))$ and by (1), $G(z)$ divides $w(z)$, implying $\deg G(z) < (\delta-1)/2$. For the first such i, (5) implies that

$$s_i(z)\,z^{\delta-1} + t_i(z)\,S(z) = r_i(z)$$

i.e.

$$r_i(z) \equiv t_i(z) S(z) \pmod{z^{\delta-1}}.$$

If $\deg r_{i-1}(z) \geq (\delta-1)/2$, then by (4),

$$\deg t_i(z) \leq \delta - 1 - \frac{(\delta-1)}{2} = \frac{\delta-1}{2}$$

and this degree is an integer. This implies that $w'(z) = r_i(z)$, $\sigma'(z) = t_i(z)$ satisfies (2). Hence $\sigma(z) = \lambda t_i(z)$ for some $\lambda \in F$ and we have deduced a scalar multiple of the error locator polynomial. By finding the roots of $t_i(z)$ we obtain the roots of $\sigma(z)$, and can deduce the error locations. Knowing the error locations, the magnitudes of the errors can be determined by solving a system of linear equations over F; this technique is illustrated in example 16 below, and discussed further in §7.4 (For an alternate method see exercise 30.). If $F = GF(2)$ then solving for the error magnitudes is not necessary, as we will see in example 15 below. These examples illustrate this general technique for decoding BCH codes.

Example 15.

Consider the polynomial $g(x) = 1 + x + x^2 + x^4 + x^5 + x^8 + x^{10}$ which generates a designed-distance 7 binary BCH code of length 15. If $GF(2^4)$ is generated by a root α of $f(x) = 1 + x + x^4$, then

$$m_1(x) = 1 + x + x^4$$

$$m_3(x) = 1 + x + x^2 + x^3 + x^4$$

$$m_5(x) = 1 + x + x^2$$

and

$$g(x) = m_1(x)\, m_3(x)\, m_5(x).$$

Now α is a primitive 15th root of unity, and α^1, α^2, α^3, α^4, α^5 and α^6 are among the roots of $g(x)$. Suppose the decoder receives

$$\mathbf{r} = (10011\ 11110\ 00110).$$

We first compute

$$r(x) = (1+x+x^6+x^8+x^9)\, m_1(x) + (x^2 + x^3)$$

$$r(x) = (x^2+x^7+x^9)\, m_3(x) + (1 + x^2)$$

$$r(x) = (1+x^2+x^5+x^8+x^9+x^{11})\, m_5(x) + x.$$

Now, using the Zech's log table for this field (see Appendix D), and the appropriate equations above,

$$S_0 = r(\alpha^1) = \alpha^2 + \alpha^3 = \alpha^6$$

$$S_1 = r(\alpha^2) = (r(\alpha))^2 = \alpha^{12}$$

$$S_2 = r(\alpha^3) = 1+\alpha^6 = \alpha^{13}$$

$$S_3 = r(\alpha^4) = (r(\alpha))^4 = \alpha^9$$

$$S_4 = r(\alpha^5) = \alpha^5$$

$$S_5 = r(\alpha^6) = (r(\alpha^3))^2 = \alpha^{11}.$$

Therefore

$$S(z) = \alpha^6 + \alpha^{12}z + \alpha^{13}z^2 + \alpha^9 z^3 + \alpha^5 z^4 + \alpha^{11}z^5.$$

Note that the coefficients in $S(z)$ are elements of $GF(2^4)$. In order to find the error locator polynomial $\sigma(z)$, we apply the extended Euclidean algorithm to $z^{\delta-1} = z^6$ and $S(z)$. Rather than writing the polynomials, we list only the associated coefficient vectors in the following table.

s_i	t_i	r_i	$\deg(r_i)$
1	0	z^6	6
0	1	$(\alpha^6,\alpha^{12},\alpha^{13},\alpha^9,\alpha^5,\alpha^{11})$	5
1	(α^{13},α^4)	$(\alpha^4,0,\alpha^6,\alpha^{12},\alpha^8)$	4
(α^2,α^3)	$(0,\alpha^{11},\alpha^7)$	$(0,\alpha^2,\alpha^3,\alpha^{14})$	3
$(1,\alpha^{11},\alpha^{12})$	$(\alpha^{13},\alpha^4,\alpha^5,\alpha)$	$(\alpha^4,0,\alpha)$	2

We stop when $\deg(r_i) < (\delta-1)/2 = 3$. The error locator polynomial is a field multiple of

$$\sigma'(z) = \alpha^{13} + \alpha^4 z + \alpha^5 z^2 + \alpha z^3.$$

The roots of $\sigma'(z)$ are the same as the roots of $\sigma(z)$. We can find the roots of $\sigma'(z)$ by trying all non-zero elements in $GF(2^4)$. A more efficient technique for finding the roots of quadratic, cubic and quartic equations will be given in §6.5. But here, by trial, we find α^2, α^{14} and α^{11} are roots of $\sigma'(z)$, and hence are the roots of $\sigma(z)$. The error locations are given by the reciprocals of these roots. They are α^{13}, α^1 and α^4. The error vector is thus

$$\mathbf{e} = (01001\ 00000\ 00010).$$

We decode $\mathbf{r}$ to $\mathbf{c} = (11010\ 11110\ 00100)$.

In general, for α a primitive nth root of unity, the set of roots of a generator polynomial $g(x)$ for a BCH code of length n and designed distance δ contains $\delta-1$ consecutive powers of the form $\alpha^a, \alpha^{a+1}, \alpha^{a+2}, \ldots, \alpha^{a+\delta-2}$. The preceding discussion considers the case $a = 1$ (i.e. narrow-sense codes). In the general case with a not necessarily equal to 1, our previous equation yielding S_i becomes

$$S_i = s(\alpha^{a+i}) = \sum_{j=0}^{l-1} y_j u_j^{a+i}, \quad 0 \le i \le \delta-2 \tag{6}$$

and we now define

$$w(z) = \sum_{i=0}^{l-1} y_i u_i^a \prod_{\substack{j=0 \\ j\neq i}}^{l-1} (1 - u_j z).$$

The theory now remains as before. This generalization is illustrated in the following example.

Example 16.

Consider a BCH code C of length 13 and designed distance $\delta = 5$ over $GF(3)$. Take the finite field $F = GF(3^3)$ generated by a root α of $f(x) = 1 + 2x^2 + x^3$ (as given in Appendix D). Then $\beta = \alpha^2$ is a primitive 13th root of unity in F. The cyclotomic cosets of 3 modulo 13 are

$$C_0 = \{0\}$$

$$C_1 = \{1,3,9\}$$

$$C_2 = \{2,5,6\}$$

$$C_4 = \{4,10,12\}$$

$$C_7 = \{7,8,11\}.$$

If $m_i(x)$ denotes the minimal polynomial of β^i, then

$$m_0(x) = x - 1$$

$$m_1(x) = 2 + 2x + 2x^2 + x^3$$

$$m_2(x) = 2 + 2x + x^3$$

$$m_4(x) = 2 + x + 2x^2 + x^3$$

$$m_7(x) = 2 + x^2 + x^3.$$

As a generator polynomial for C we may take

$$g(x) = m_0(x)\, m_1(x)\, m_2(x) = 2 + 2x + x^4 + 2x^5 + x^6 + x^7.$$

We can prove that C actually has distance $d = 6$ as follows. Since $x - 1$ divides $g(x)$, it divides all $c(x) \in C$, and hence $c(1) = 0$. Since the non-zero terms of $c(x)$ have coefficients 1 and -1 and there must be at least 5 such terms (since $\delta = 5$), it follows that $\mathrm{wt}(c(x)) \geq 6$ for $c(x) \neq 0$. Hence $d \geq 6$, and since $\mathrm{wt}(g(x)) = 6$, the distance of C is exactly 6.

An alternate approach to establishing d is using the techniques of §6.3. Let $c(x) \in C$ and define $S = \{\rho\colon c(\rho) = 0\}$. Clearly $C_0 \cup C_1 \cup C_2 \subseteq S$. If S contains elements from both C_4 and C_7, then it contains all elements of both C_4 and C_7 and $c(x) = 0$. If S contains elements from one but not both of C_4 and C_7, then it contains all the elements from that coset and none from the other, and it is easily proven (as in example 14) that $c(x)$ has weight at least 6. Hence, assume S contains no elements from C_4 or C_7. Define the following sequence $(A_i\colon i \geq 0)$ of independent sets where $A_0 = \varnothing$, $A_{i+1} = c_i A_i \cup \{b_i\}$ and $b_i \notin S$. (For simplicity, we write only the exponent i of α^i.)

$A_0 = \varnothing$		$b_0 = 4$
$A_1 = \{4\}$	$c_1 = -3$	$b_1 = 10$
$A_2 = \{1,10\}$	$c_2 = -1$	$b_2 = 10$
$A_3 = \{0,9,10\}$	$c_3 = -4$	$b_3 = 4$
$A_4 = \{5,6,9,4\}$	$c_4 = -4$	$b_4 = 8$
$A_5 = \{1,2,5,0,8\}$	$c_5 = 1$	$b_5 = 10$
$A_6 = \{1,2,3,6,9,10\}$		

Since $|A_6|$, we have $\mathrm{wt}(c(x)) \geq 6$ (by Theorem 6.3), and since $\mathrm{wt}(g(x)) = 6$, it follows that C has distance exactly 6.

Now suppose the decoder receives

$$\mathbf{r} = (220\,021\,110\,2110).$$

First, we divide $r(x)$ by $m_0(x)$, $m_1(x)$ and $m_2(x)$.

$$r(x) = q_0(x)\, m_0(x) + 1$$

$$r(x) = q_1(x)\,m_1(x) + (1 + 2x + x^2)$$

$$r(x) = q_2(x)\,m_2(x) + (1 + 2x).$$

The generalized equation (6) with $a = 0$ gives $S_i = r(\beta^i)$. Using the Zech's log table for $GF(3^3)$ in Appendix D,

$$S_0 = r(\beta^0) = 1$$

$$S_1 = r(\beta^1) = 1 + 2\beta + \beta^2 = 1 + 2\alpha^2 + \alpha^4 = \alpha^{14}$$

$$S_2 = r(\beta^2) = 1 + 2\beta^2 = 1 + 2\alpha^4 = \alpha^{23}$$

$$S_3 = r(\beta^3) = (r(\beta))^3 = (\alpha^{14})^3 = \alpha^{42} = \alpha^{16}.$$

This gives

$$S(z) = 1 + \alpha^{14}z + \alpha^{23}z^2 + \alpha^{16}z^3.$$

We now apply the Euclidean algorithm to $z^{\delta-1} = z^4$ and $S(z)$.

s_i	t_i	r_i	$\deg(r_i)$
1	0	z^4	4
0	1	$(1,\alpha^{14},\alpha^{23},\alpha^{16})$	3
1	$(\alpha^{17},\alpha^{23})$	$(\alpha^{17},\alpha^{16},\alpha^{13})$	2
$-(\alpha^{16},\alpha^3)$	$(\alpha^{15},\alpha^3,\alpha^{13})$	$(\alpha^{15},\alpha^{16})$	1

We continue until $\deg(r_i) < (\delta-1)/2 = 2$. At this point, we know that the error locator polynomial is a scalar multiple of

$$\sigma'(z) = \alpha^{15} + \alpha^3 z + 2z^2.$$

Over $GF(3)$, we can apply the standard quadratic formula to find the roots of this quadratic equation. We get

$$z = \frac{-\alpha^3 \pm \sqrt{(\alpha^3)^2 - 4(2)\alpha^{15}}}{2(2)}$$

$$= -\alpha^3 \pm \sqrt{\alpha^6 + \alpha^{15}} = -\alpha^3 \pm \sqrt{\alpha^{12}} = -\alpha^3 \pm \alpha^6.$$

Therefore the two roots are

$$-\alpha^3+\alpha^6=\alpha^{10}=\beta^5 \quad \text{and} \quad -\alpha^3-\alpha^6=\alpha^{18}=\beta^9.$$

The error locations are thus $\beta^{-5}=\beta^8$ and $\beta^{-9}=\beta^4$. Let $e(x)=y_1x^4+y_2x^8$. To determine the error magnitudes y_1 and y_2, which are elements of $GF(3)$, we proceed as follows. Note

$$S_1=r(\beta^1)=e(\beta^1)=y_1\beta^4+y_2\beta^8$$

$$S_2=r(\beta^2)=e(\beta^2)=y_1(\beta^4)^2+y_2(\beta^8)^2$$

This yields the linear system over $GF(3^3)$

$$\begin{bmatrix}\alpha^{14}\\ \alpha^{23}\end{bmatrix}=\begin{bmatrix}\beta^4 & \beta^8\\ \beta^8 & \beta^{16}\end{bmatrix}\begin{bmatrix}y_1\\ y_2\end{bmatrix}.$$

Hence

$$\begin{bmatrix}\alpha^4 & \alpha\\ \alpha & \alpha^6\end{bmatrix}\begin{bmatrix}\alpha^{14}\\ \alpha^{23}\end{bmatrix}=\begin{bmatrix}y_1\\ y_2\end{bmatrix}$$

giving $y_1=2$ and $y_2=2$. The error vector is

$$\mathbf{e}=(000\ 020\ 002\ 0000)$$

and we decode $\mathbf{r}$ to $\mathbf{c}=\mathbf{r}-\mathbf{e}=(220\ 001\ 111\ 2110)$.

6.5 Linearized Polynomials and Finding Roots of Polynomials†

The previous section indicates a need to find roots of polynomials over finite fields. The method of trying all possibilities (e.g. in example 15) will always work, but this is not very efficient if the field is large. In what follows we describe an efficient method for finding roots of polynomials over $GF(2^n)$ for polynomials of degree 2, 3 or 4.

Definition. Let $F=GF(p^n)$, where p is a prime. An *affine polynomial* over F is any polynomial $l(x)$ of the form

$$l(x)=\lambda+\sum_{i=0}^{t}\lambda_i x^{p^i}=\lambda+\lambda_0x+\lambda_1x^p+\lambda_2x^{p^2}+\cdots+\lambda_tx^{p^t}$$

† This section may be omitted without loss of continuity.

where $\lambda, \lambda_i \in F, 0 \le i \le t$. If $\lambda = 0$ then $l(x)$ is a *linearized polynomial.*

Example 17.

The polynomial $l(x) = 1 + x^2 + x^8$ is an affine polynomial over $GF(2)$.

Finding the roots of an affine polynomial is relatively easy. It involves solving a linear system of equations over F, as illustrated by the following example.

Example 18.

Suppose we wish to find the roots of $l(x) = 1 + x^2 + x^8$ in $F = GF(2^4)$. We may take F to be generated by α, where $f(x) = 1 + x + x^4$ and $f(\alpha) = 0$. The elements of F are 4-tuples over $GF(2)$, with respect to the polynomial basis $B = \{1, \alpha, \alpha^2, \alpha^3\}$. Suppose $\rho = (a_0a_1a_2a_3)$ is a root of $l(x)$, $a_i \in GF(2)$. Then $\rho = \sum_{i=0}^{3} a_i\alpha^i$ where $\alpha = (0100)$, and since $l(\rho) = 0$,

$$1 + (\sum_{i=0}^{3} a_i\alpha^i)^2 + (\sum_{i=0}^{3} a_i\alpha^i)^8 = 0$$

and, using Lemma 2.7,

$$1 + \sum_{i=0}^{3} a_i\alpha^{2i} + \sum_{i=0}^{3} a_i\alpha^{8i} = 0.$$

Expanding this, noting $\alpha^{16} = \alpha$, and collecting terms gives

$$1 + a_1(\alpha^2+\alpha^8) + a_2(\alpha^4+\alpha) + a_3(\alpha^6 + \alpha^9) = 0.$$

Using F as given in Appendix D, $\alpha^2 + \alpha^8 = (1000)$, $\alpha^4 + \alpha = (1000)$ and $\alpha^6 + \alpha^9 = (0110)$. Hence we obtain

$$(1+a_1+a_2, a_3, a_3, 0) = (0, 0, 0, 0)$$

and the linear system

$$1 + a_1 + a_2 = 0$$

$$a_3 = 0.$$

There are 4 solutions to this system: (0010), (0100), (1010), (1100), i.e. α^2, α, α^8 and α^4. We could have observed this right from the beginning, since $l(x) = 1 + x^2 + x^8 = (1+x+x^4)^2$, but the method given here is more general.

Example 19.

Consider finding the roots of $l(x) = 1 + \alpha^3 x + \alpha^5 x^8$ over $F = GF(2^4)$, generated by a root α of $f(x)$ where $f(x) = 1 + x + x^4$. Let $\rho = (a_0 a_1 a_2 a_3)$ be a root of $l(x)$. Then

$$\rho = a_0 + a_1\alpha + a_2\alpha^2 + a_3\alpha^3$$

and $l(\rho) = 0$. Hence

$$\begin{aligned} l(\rho) &= 1 + \alpha^3(a_0 + a_1\alpha + a_2\alpha^2 + a_3\alpha^3) + \alpha^5(a_0 + a_1\alpha^8 + a_2\alpha + a_3\alpha^9) \\ &= 1 + a_0(\alpha^3 + \alpha^5) + a_1(\alpha^4 + \alpha^{13}) + a_2(\alpha^5 + \alpha^6) + a_3(\alpha^6 + \alpha^{14}) \\ &= 1 + a_0(0111) + a_1(0111) + a_2(0101) + a_3(1010) \\ &= (1{+}a_3,\ a_0{+}a_1{+}a_2,\ a_0{+}a_1{+}a_3,\ a_0{+}a_1{+}a_2) \\ &= (0, 0, 0, 0) \end{aligned}$$

and we obtain the system of linear equations

$$1 + a_3 = 0$$

$$a_0 + a_1 + a_2 = 0$$

$$a_0 + a_1 + a_3 = 0$$

$$a_0 + a_1 + a_2 = 0$$

which has solutions $(a_0 a_1 a_2 a_3) = (0111)$ and (1011). Therefore, the only roots of $l(x)$ in F are α^{11} and α^{13}.

It is clear from the definition that every quadratic polynomial over $GF(2^n)$ is an affine polynomial. Unfortunately, not all non-quadratic polynomials are affine, and hence the above technique for finding roots cannot always be applied. However, in certain cases, it is possible to transform a non-affine polynomial into an affine polynomial, find the roots of this transformed polynomial (by solving the appropriate system of linear equations), and then from these recover the roots of the original polynomial. We consider below the particular cases of polynomials of degree 3 and 4 over $GF(2^n)$.

Consider the cubic polynomial

$$h(x) = ax^3 + bx^2 + cx + d$$

where $a, b, c, d \in GF(2^n)$, $a \neq 0$. If $d = 0$, then $h(x)$ has 0 as a root and the problem reduces to that of finding the roots of a quadratic polynomial over $GF(2^n)$. If $d \neq 0$, let $x = y + b/a$. Then

$$\begin{aligned} h(x) &= ax^3 + bx^2 + cx + d \\ &= ay^3 + (\frac{b^2}{a} + c)y + (\frac{cb}{a} + d) \\ &= \tilde{h}(y). \end{aligned}$$

and $y\tilde{h}(y)$ is a linearized polynomial. The non-zero roots of $y\tilde{h}(y)$ are the roots of $\tilde{h}(y)$ and these will give the roots of $h(x)$. Hence we have the following result.

Theorem 6.5. Any cubic polynomial over $GF(2^n)$ can be transformed into a linearized polynomial from which its roots in $GF(2^n)$ can be deduced.

Example 20.

Suppose we wish to find the roots of $h(x) = \alpha^{13} + \alpha^4 x + \alpha^5 x^2 + \alpha x^3$ over $GF(2^4)$, where $GF(2^4)$ is generated by a root α of $f(x) = 1 + x + x^4$. Let $x = y + \alpha^4$. Then

$$\begin{aligned} \tilde{h}(y) &= \alpha(y + \alpha^4)^3 + \alpha^5(y + \alpha^4)^2 + \alpha^4(y + \alpha^4) + \alpha^{13} \\ &= \alpha y^3 + \alpha^{14} y + \alpha^3 \end{aligned}$$

and

$$l(y) = y\tilde{h}(y) = \alpha y^4 + \alpha^{14} y^2 + \alpha^3 y$$

is a linearized polynomial. Let $\rho = (a_0 a_1 a_2 a_3)$ be a root of $l(y)$. Then

$$\begin{aligned} l(\rho) &= \alpha(\sum a_i \alpha^i)^4 + \alpha^{14}(\sum a_i \alpha^i)^2 + \alpha^3(\sum a_i \alpha^i) \\ &= a_0(1 + \alpha) + a_1(1 + \alpha + \alpha^2) + a_2(\alpha^2) + a_3(1 + \alpha + \alpha^2) \\ &= 0. \end{aligned}$$

Equating coefficients to 0 gives the linear system

$$a_0 + a_1 + a_3 = 0$$

$$a_0 + a_1 + a_3 = 0$$

$$a_1 + a_2 + a_3 = 0.$$

There are four solutions to this system, and

$$(a_0a_1a_2a_3) \in \{(0000), (1110), (0101), (1011)\}.$$

So the roots of $y\tilde{h}(y)$ are $0, \alpha^{10}, \alpha^9, \alpha^{13}$. The roots of $\tilde{h}(y)$ are thus $\alpha^{10}, \alpha^9, \alpha^{13}$, and since $x = y + \alpha^4$, the roots of $h(x)$ are α^2, α^{14} and α^{11}.

For quartic equations, we have a similar situation.

Theorem 6.6. Any quartic equation over $GF(2^n)$ can be transformed into an affine polynomial from which its roots in $GF(2^n)$ can be deduced.

To see this, consider the quartic equation

$$h(x) = ax^4 + bx^3 + cx^2 + dx + e$$

where $a, b, c, d, e \in F$, $a \neq 0$. We assume $e \neq 0$ for otherwise $h(x) = x\,c(x)$ where $c(x)$ is a cubic equation. We also assume that $b \neq 0$ (since if $b=0$, then $h(x)$ is already an affine polynomial). With $b \neq 0$, let $x = y+t$ and $\tilde{h}(y) = h(y+t)$, where $t^2 = d/b$. Since we are working in $GF(2^n)$, d/b has a unique square root and t is uniquely defined (see below). We see that

$$\tilde{h}(y) = ay^4 + by^3 + \tilde{c}y^2 + \tilde{d}$$

where

$$\tilde{c} = bt + c$$

$$\tilde{d} = at^4 + bt^3 + ct^2 + dt + e.$$

Now $\tilde{h}(y)$ is not an affine polynomial, nor is $y \cdot \tilde{h}(y)$. But suppose we let $y = 1/z$. Then

$$\tilde{h}(\frac{1}{z}) = a\frac{1}{z^4} + b\frac{1}{z^3} + \tilde{c}\frac{1}{z^2} + \tilde{d}$$

and

$$l(z) = z^4 \cdot \tilde{h}(\frac{1}{z}) = a + bz + \tilde{c}z^2 + \tilde{d}z^4$$

is an affine polynomial in the variable z. The reciprocals of the roots of $l(z)$ are the roots of $\tilde{h}(y)$, from which the roots of $h(x)$ can be deduced.

To determine the unique square root of d/b in $GF(2^n)$, we may proceed as follows. Suppose $d/b = \alpha^j$, where α is a primitive element in $GF(2^n)$. Letting $\alpha^{2d} = \alpha^j$, we wish to find $\sqrt{\alpha^j} = \alpha^d$, i.e. we wish to find d such that

$$2d \equiv j \pmod{2^n - 1} \quad \text{or} \quad d \equiv 2^{-1} \cdot j \pmod{2^n - 1}.$$

This is available directly upon computing $2^{-1} \pmod{2^n-1}$ via the extended Euclidean algorithm (see Appendix B).

Example 21.

Consider finding the roots of $h(x) = \alpha^8 + \alpha^4 x + \alpha x^2 + \alpha^8 x^3 + x^4$ in $GF(2^4)$, generated by α where $f(x) = 1 + x + x^4$ and $f(\alpha) = 0$. Let $x = y+t$, where $t^2 = \alpha^{11}$, implying $t = \alpha^{13}$. Then

$$\tilde{h}(y) = y^4 + \alpha^8 y^3 + \alpha^{11} y^2 + 1.$$

Setting $y = \frac{1}{z}$ we get

$$l(z) = 1 + \alpha^8 z + \alpha^{11} z^2 + z^4$$

which is an affine polynomial in z. Let

$$\rho = (a_0 a_1 a_2 a_3) = a_0 + a_1\alpha + a_2\alpha^2 + a_3\alpha^3$$

be a root of $l(z)$. Then

$$\begin{aligned} 0 &= l(\rho) \\ &= 1 + a_0(\alpha^8+\alpha^{11}+1) + a_1(\alpha^9+\alpha^{13}+\alpha^4) + a_2(\alpha^{10}+1+\alpha^8) + a_3(\alpha^{11}+\alpha^2+\alpha^{12}) \\ &= 1 + a_0\alpha^9 + a_1\alpha^2 + a_2\alpha^4 + a_3\alpha^8, \end{aligned}$$

using the field representation and Zech's log table as given in Appendix D. This gives the linear system

$$1 + a_2 + a_3 = 0$$

$$a_0 + a_2 = 0$$

$$a_1 + a_3 = 0$$

$$a_0 = 0.$$

The only solution to this system is (0101). Hence $z = \alpha^9$ is a root of $l(z)$, and $y = \alpha^6$ is a root of $\bar{h}(y)$. This gives $x = \alpha^6 + \alpha^{13} = 1$ as a root of the original equation, $h(x)$.

6.6 Exercises

General.

1. In example 8 of this chapter, do we get the same polynomial $g(x)$ if a different irreducible polynomial is used for the representation of $GF(2^4)$?

2. Give an alternate proof of Theorem 6.2, using the idea of the proof of Theorem 6.4.

3. For $\sigma(z)$ and $w(z)$ as defined in §6.4, prove that $\gcd(w(z), \sigma(z)) = 1$.

4. For the codes of examples 15 and 16 of this chapter, describe briefly how decoding schemes from previous chapters could be used.

BCH Codes.

5. (a) Give a generator polynomial for a double-error correcting BCH code C of length 8 and dimension 3 over $GF(3)$.

   (b) Construct a parity-check matrix H for C.

6. Consider $GF(2^4)$ generated by a root α of $f(x) = 1 + x + x^4$. The polynomial $g(x) = m_{\alpha^1}(x)\, m_{\alpha^3}(x)\, m_{\alpha^5}(x) = 1 + x + x^2 + x^4 + x^5 + x^8 + x^{10}$ generates a binary cyclic (15,5)-code with distance 7. Consider the received vector $\mathbf{r} = (110\;000\;111\;000\;000)$. Given that

$$r(x) = q_1(x)\, m_{\alpha^1}(x) + 1$$

$$r(x) = q_2(x)\, m_{\alpha^3}(x) + 1 + x^2 + x^3$$

$$r(x) = q_3(x)\, m_{\alpha^5}(x) + x^2$$

   (a) Find the syndrome polynomial $S(z)$.

   (b) Find the error locator polynomial $\sigma(z)$.

   (c) Find the roots of $\sigma(z)$.

   (d) Determine the error pattern associated with $\mathbf{r}$.

7. Note the list of minimum polynomials given with $GF(2^5)$ in Appendix D.

   (a) Construct a generator polynomial for a (31,11)-binary cyclic code which has designed distance 11.

   (b) Construct a generator polynomial $g(x)$ for a (31,15)-binary cyclic code which has designed distance 8 and is self-orthogonal.

(c) Determine the number of cyclic self-orthogonal codes of length 31 in $V_{31}(Z_2)$.

8. Let $g(x) = (1+x+x^2)(1+x+x^4)(1+x+x^2+x^3+x^4)$ be the generator polynomial for a (15,5)-BCH code with minimum distance 7. Computations are done in $GF(16)$ which is generated by a root α of $f(x)=x^4+x+1$. Determine, if possible, the error positions in the following received vectors.

   (a) $\mathbf{r}_1 = (100\ 111\ 001\ 011\ 101)$ (b) $\mathbf{r}_2 = (111\ 111\ 000\ 101\ 101)$

   (c) $\mathbf{r}_3 = (000\ 111\ 111\ 111\ 000)$

9. Construct a generator polynomial $g(x)$ for a (15,7)-BCH code over $GF(4)$ with designed distance 7.

10. Show that the extended t-error-correcting binary BCH code formed by adding an overall parity-check to the end of a cyclic binary BCH code is capable of correcting all patterns of t or fewer errors and all bursts of $t+1$ consecutive errors which do not include the overall parity-check position (cf. exercise 42, Chapter 5).

11. Prove that if $g(x)$ generates a cyclic (n,k)-code over $GF(q)$ having distance d, then $g(x^t)$ generates an (nt,kt)-code over $GF(q)$ having distance d.

12. Let α be a generating element for $GF(2^m)$ and let $g(x) = lcm\{m_{\alpha^i}(x)\text{: } 1 \le i \le 2t\}$, i.e. $g(x)$ generates a binary BCH code C of length $n = 2^m - 1$ having designed distance $\delta = 2t+1$. Prove that if δ divides n then C has distance $d = \delta$.

13. Consider $F = GF(8)$ defined using the polynomial $f(x) = 1 + x + x^3$.

   (a) Construct a Zech's log table for this field, using a generator α where $f(\alpha) = 0$.

   (b) Construct a generator polynomial $g(x)$ for a (7,3)-code C over F having designed distance 5. (Note that C consists of 7-tuples over $GF(8)$ [not $GF(2)$], and that the minimum polynomial of an element of $GF(8)$ over $GF(8)$ is always linear.)

   (c) Suppose that in each codeword we replace each component by the binary 3-tuple which represents the field element of $GF(8)$ in that position. The result is a binary vector of length 21. Let C' be the set of all binary 21-tuples obtained from the codewords of C. Show that C' is a binary (21,9)-code.

   (d) Prove that C' will correct up to 2 errors and will correct all burst errors of length 4 or less.

Linearized polynomials and finding roots.

14. Find the roots of $g(x) = \alpha^4 + \alpha x^3 + x^7$ in $GF(2^5)$, where $GF(2^5)$ is generated by a root α of $f(x) = 1 + x^2 + x^5$.

15. Find the roots of $g(x) = \alpha^5 + x + \alpha^2 x^3 + x^7$ in the field of the problem above.

16. Show that any cubic polynomial over $GF(3^n)$ can be transformed to a linearized polynomial.

17. Find the roots in $GF(3^3)$ of each of the following polynomials, where $GF(3^3)$ is generated by a root α of $f(x) = 1 + 2x^2 + x^3$.

    (a) $a(x) = x^3 + \alpha^{21}x + \alpha^{21}$ (b) $b(x) = x^2 + \alpha^{10}x + \alpha^4$

    (c) $c(x) = x^3 + 2\alpha^{20}x^2 + \alpha^{13}x + 2\alpha^7$

18. Find the roots in $GF(5^2)$ of each of the following polynomials, where $GF(5^2)$ is generated by $f(x) = 2 + x + x^2$ and $f(\alpha) = 0$:

    (a) $a(x) = 4 + 3\alpha^2 x + \alpha x^2$ (b) $b(x) = \alpha^3 + 2x + \alpha x^5$ (c) $c(x) = \alpha^5 + 4x^4 + x^{24}$

19. Find the roots in $GF(3^4)$ of each of the following polynomials, where $GF(3^4)$ is generated by $f(x) = x^4 + x + 2$ and $f(\alpha) = 0$.

    (a) $a(x) = 1 + 2\alpha^3 x + \alpha x^2$ (b) $b(x) = \alpha^{11} + 2\alpha^7 x + \alpha x^2 + x^3$

    (c) $c(x) = \alpha^9 + 2x^2 + \alpha^2 x^3$ (d) $d(x) = 1 + x^2 + x^8$

General.

20. Let

$$H = \begin{bmatrix} \alpha^0 & \alpha^2 & \alpha^4 & \alpha^6 & \cdots & \alpha^{20} & \alpha^{22} & \alpha^{24} \\ \alpha^0 & \alpha^{-2} & \alpha^{-4} & \alpha^{-6} & \cdots & \alpha^{-20} & \alpha^{-22} & \alpha^{-24} \end{bmatrix}$$

be a parity-check matrix for a $(13,k)$-code C over Z_3. (H is a 6×13 matrix over Z_3 when α^i is replaced by the corresponding column 3-tuple.) Note the Zech's log table given for the field $GF(3^3)$ in Appendix D.

    (a) Determine whether or not C is a cyclic code. If C is cyclic, determine the generator polynomial.

    (b) Determine k.

    (c) Prove that C will correct up to 2 errors.

(d) Find the error pattern associated with the vector $\mathbf{r} = (1121\ 0101\ 1010\ 0)$.

21. Construct a generator polynomial $g(x)$ for a (20,6)-cyclic code C over $GF(3)$ so that C has distance $d \geq 10$.

22. Construct a generator matrix for a 2 error-correcting (13,7)-code containing 2187 codewords. Describe a decoding procedure for this code.

23. Construct a generator polynomial for a single error correcting (9,3)-code.

24. A code is said to be *reversible* if and only if $(c_0c_1...c_{n-1})$ is a codeword whenever $(c_{n-1}c_{n-2}...c_1c_0)$ is.

    (a) Prove that a cyclic code is reversible if and only if the reciprocal of every root of its generator polynomial is also a root of the generator polynomial.

    (b) Prove that if -1 is a power of q mod n then every cyclic code of length n over $GF(q)$ is reversible.

25. Construct a generator polynomial for a cyclic code C of block length 14 and dimension as large as possible over $GF(2)$ having distance at least 4.

26. Let α be a primitive 21st root of unity in $GF(2^6)$. Then α is a root of $x^{21}-1$. Let $g(x)$ be a divisor of $x^{21}-1$ such that for some integer b, $\alpha^b, \alpha^{b+1}, ..., \alpha^{b+d-2}$ are roots of $g(x)$.

    (a) Prove that $g(x)$ generates a cyclic subspace having minimum distance at least d.

    (b) Using the result of (a), deduce that at least two of the cyclic subspaces of dimension 11 in V_{21} have minimum distance at least 5.

27. Prove the properties of the extended Euclidean algorithm given by equations (4) and (5) in §6.4.

28. Prove that the binary Hamming codes are (equivalent to) cyclic codes.

29. Let $g(x)$ be a polynomial over $GF(q)$, and let its distinct irreducible factors over $GF(q)$ be $g_1(x), ..., g_t(x)$, with respective degrees $n_1, \ldots, n_t$. Prove that the splitting field $GF(q^m)$ of $g(x)$ is determined by the integer $m = lcm(n_1, \ldots, n_t)$.

30. Using the notation of §6.4 show that the error magnitudes are given by

$$y_k = \frac{u_k^{-1}\, w(u_k^{-1})}{\prod_{j=0}^{l-1} (1-u_j u_k^{-1})} = \frac{-w(u_k^{-1})}{\sigma(u_k^{-1})}$$

where $w(z)$ is the error evaluator polynomial and $\sigma(z)$ is the error locator polynomial.

31. Find the roots of $g(x) = \alpha^{19} + x + \alpha^{22}x^2 + \alpha^{22}x^3 + x^4$ in $GF(2^5)$ where $GF(2^5)$ is generated by a root α of $f(x) = 1 + x^2 + x^5$.

32. Find the roots of $g(x) = \alpha + \alpha^9 x + \alpha^7 x^2 + \alpha^{14} x^3 + x^4$ in $GF(2^4)$ where $GF(2^4)$ is generated by a root of $f(x) = 1 + x + x^4$.

Chapter 7

ERROR CORRECTION TECHNIQUES and DIGITAL AUDIO RECORDING

7.1 Introduction

In this chapter, we consider a special class of BCH codes and several important techniques available to enhance error correction capabilities. In particular, we introduce the well known Reed-Solomon codes, and discuss the ideas of channel erasures and interleaving. While these concepts are of general interest, we find it motivating to focus our discussion, bringing these ideas together by considering their particular application in the digital audio recording industry.

7.2 Reed-Solomon Codes

Let β be an element of $F = GF(q)$ and suppose that the order of β is n (i.e. $\beta^n = 1$ but $\beta^s \neq 1$ for any positive $s < n$). Let

$$g(x) = (x-\beta^{1+a})(x-\beta^{2+a})\cdots(x-\beta^{\delta-1+a})$$

for some $\delta \geq 2$ and some $a \geq 0$. Then since $g(x)$ divides $x^n - 1$, $g(x)$ is the generator polynomial for an ideal in $F[x]/(x^n-1)$, and hence generates a cyclic code in $V_n(F)$. Since $\beta^i \in F$ for any i, the minimum polynomial of β^i is simply $x - \beta^i$. It follows from the definition of BCH codes that $g(x)$ generates a BCH code with designed distance δ. In this situation the code generated by $g(x)$ is called a *Reed-Solomon code,* and is an (n,k)-code over F with $n-k = \delta-1$, i.e. $k = n-\delta+1$. The difference here is that whereas for BCH codes in general β is an element of some extension field of F, for Reed-Solomon codes β is actually an element of F itself. We will refer to such a code as an *(n,k)-RS code* over F. Applying the BCH bound on the distance d of the code, we get that $d \geq \delta$. In fact, we will shortly prove that $d = \delta$. Before doing so, we consider several examples of Reed-Solomon codes.

Example 1.

Consider $GF(2^4)$ as given in Appendix D, generated by a root α of $f(x) = 1 + x + x^4$. The element $\beta = \alpha^3$ is a primitive 5th root of unity. Let

$$g(x) = (x-\beta)(x-\beta^2)(x-\beta^3) = \alpha^3 + \alpha^2 x + \alpha^{11} x^2 + x^3.$$

Then $g(x)$ generates a (5,2)-RS code C over $GF(2^4)$ with designed distance $\delta = 4$. A generator matrix for C is

$$G = \begin{bmatrix} \alpha^3 & \alpha^2 & \alpha^{11} & 1 & 0 \\ 0 & \alpha^3 & \alpha^2 & \alpha^{11} & 1 \end{bmatrix}$$

and a parity check matrix is

$$H = \begin{bmatrix} 1 & 0 & 0 & \alpha^3 & \alpha^{14} \\ 0 & 1 & 0 & \alpha^2 & \alpha^8 \\ 0 & 0 & 1 & \alpha^{11} & \alpha^{12} \end{bmatrix}.$$

C has 256 codewords. That the distance of C is actually $d = 4$ could be checked by generating all codewords and examining their weights.

Example 2.

Consider the finite field $GF(5)$ and let $\alpha = 2$. It is easily checked that $ord(\alpha) = 4$, and $\beta = \alpha$ is thus a primitive 4th root of unity. Let

$$g(x) = (x-2)(x-4) = 3 + 4x + x^2.$$

Then $g(x)$ generates a (4,2)-RS code C over $GF(5)$. A generator matrix for C is

$$G = \begin{bmatrix} 3 & 4 & 1 & 0 \\ 0 & 3 & 4 & 1 \end{bmatrix}$$

and a parity check matrix is

$$H = \begin{bmatrix} 1 & 0 & 2 & 2 \\ 0 & 1 & 1 & 3 \end{bmatrix}.$$

C has distance 3 and contains 25 codewords.

Theorem 7.1. An (n,k)-RS code C with designed distance δ has distance
$$d = n - k + 1 = \delta.$$

Proof.

Let $g(x)$ be the generating polynomial for C. They by definition, there exists an element $\alpha \in F$ of order n such that

$$g(x) = \prod_{i=1}^{\delta-1} (x - \alpha^i)$$

where $\delta = n - k + 1$. We know from Theorem 6.4 that C has minimum distance $d \geq \delta$. Since $g(x)$ has at most δ non-zero terms in its expansion and $g(x)$ is a codeword, we have that $d \leq \delta$. Hence $d = \delta$. □

Now the distance of any (n,k)-code can be shown to satisfy the *Singleton bound* $d \leq n - k + 1$ (exercise 69, Chapter 3); and linear codes which meet this bound with equality are known as *maximum-distance separable (MDS) codes* (see exercise 70, Chapter 3). Hence Reed-Solomon codes are MDS codes – and for fixed n and k, no linear code can have larger distance than a Reed-Solomon code.

Using Theorem 1.2, Theorem 7.1 leads directly to the following result.

Theorem 7.2. An (n,k)-RS code over F can detect $n - k$ errors or correct $\lfloor (n-k)/2 \rfloor$ errors.

In practice, Reed-Solomon codes over $GF(2^m)$ are of greatest interest, and we shall focus here on these. This interest is in large part due to the burst error correcting capabilities Reed-Solomon codes afford, and the ease of implementation of codes over fields of characteristic 2.

Consider an (n,k)-RS code C over $GF(2^m)$. Now each element in $GF(2^m)$ can be represented as an m-tuple over $GF(2)$. Hence each codeword $\mathbf{c} \in C$, an n-tuple over $GF(2^m)$, can be replaced by an mn-tuple $\mathbf{c}'$ over $GF(2)$ by replacing each $GF(2^m)$ component of $\mathbf{c}$ with a binary m-tuple. The resulting set of vectors C' is an (nm,km)-code over $GF(2)$. This follows easily from the observations that C' contains 2^{km} vectors and is closed under vector addition and scalar multiplication from $GF(2)$.

We now consider the ability of C' to correct a burst error $\mathbf{e}$ of bitlength b. Let $\mathbf{c}$ be any codeword in C with corresponding codeword $\mathbf{c}'$ in C'. Also, let c_i be the ith symbol in $\mathbf{c}$, and let $\mathbf{c}_i'$ be the binary m-tuple corresponding to c_i. By Theorem 7.2, C' will certainly be able to correct the error **e** if **e** alters at most $\lfloor (n-k)/2 \rfloor$ symbols in **c**. Aligning the first non-zero bit in **e** to the last bit position of $\mathbf{c}_i'$ for some i, then C' will certainly be capable of correcting **e** if the remaining $b-1$ entries in the burst do not affect more than the next $\lfloor (n-k)/2 \rfloor - 1$ symbols in **c**. Hence if $b-1 < m\,(\lfloor (n-k)/2 \rfloor - 1)$, C' can correct **e**. This leads to the following result.

Theorem 7.3. An (n,k)-RS code over $GF(2^m)$ implies the existence of a binary (nm,km)-code with burst error correcting capability $m\,(\lfloor (n-k)/2 \rfloor - 1) + 1$.

An example at this point may be helpful.

Example 3.

Let $F = GF(2^4)$ be defined by a root α of $f(x) = 1 + x + x^4$. Then $\beta = \alpha^3$ is a primitive 5th root of unity, and β has minimal polynomial $g(x) = 1 + x + x^2 + x^3 + x^4$ (see example 6, Chapter 2). Now

$$g(x) = (x-\beta)(x-\beta^2)(x-\beta^3)(x-\beta^4),$$

and hence $g(x)$ is the generator polynomial for a (5,1)-RS code C over F of designed distance $\delta = 5$. The generator matrix for C over F is $G = [1,1,1,1,1]$, and

$$C = \{(\alpha^i,\alpha^i,\alpha^i,\alpha^i,\alpha^i) : 0 \le i \le 14\} \cup \{(0,0,0,0,0)\}.$$

Each codeword $\mathbf{c} \in C$ is a 5-tuple over F. Replacing each of the 5 components of **c** by a corresponding 4-tuple over $GF(2)$ results in a code C' in which each codeword is a binary 20-tuple. Any binary burst error of length 5 or less in C' can be corrected.

7.3 Channel Erasures

A *channel erasure* is an error for which the position of the error is known but the magnitude is not. An erasure in position i of a received n-tuple can occur as a result of the decoder receiving insufficient information to decide on a symbol for that coordinate, or information indicating that a particular coordinate value is unreliable. The task of the decoder here is to restore or "fill" the erasure positions.

Let C be an $[n,M]$-code over an alphabet A. A received vector **r** having l erasures (and no additional errors) is said to be correctable to a codeword **c** if **c** is the unique codeword among the $|A|^l$ vectors which agree with **r** in the $n-l$ non-erasure positions. If there is no unique codeword in this set then **r** is not correctable. A code C is said to correct u erasures if any received vector with $l \le u$ erasures is correctable. With this, we have the following result. (Compare to

Theorem 1.2.)

Theorem 7.4 An $[n,M]$-code C of distance d, defined over an alphabet A, is capable of correcting $d-1$ erasures.

Proof.

Suppose **r** is a received n-tuple containing $l = d-1$ erasures. Among the $|A|^l$ n-tuples which agree with **r** in the $n-l$ non-erasure positions there can be at most one codeword, since the minimum distance in C is d. Assuming the channel admits only erasures, there thus is exactly one such codeword **c**, to which **r** can be corrected. □

For linear codes over $GF(q)$, erasures can be determined by solving a system of linear equations, as illustrated by the following example.

Example 4.

Consider the code given in example 1. Suppose the decoder receives the vector

$$\mathbf{r} = (\_\ \alpha^6\ \_\ \_\ 1),$$

where "_" indicates an erasure position, and arbitrarily assigns 0's to the erasure positions yielding

$$\mathbf{r}_0 = (0\ \alpha^6\ 0\ 0\ 1).$$

Then

$$\mathbf{r}_0 = \mathbf{c} + (u_1\ 0\ u_2\ u_3\ 0)$$

for some codeword **c**. Since $H\mathbf{c}^T = 0$ we have that $H\mathbf{r}_0^T = H(u_1\,0\,u_2\,u_3\,0)^T$, i.e.

$$\begin{bmatrix} \alpha^{14} \\ \alpha^{14} \\ \alpha^{12} \end{bmatrix} = \begin{bmatrix} u_1 + \alpha^3 u_3 \\ \alpha^2 u_3 \\ u_2 + \alpha^{11} u_3 \end{bmatrix}.$$

This system of 3 equations in 3 unknowns has the unique solution $u_1 = \alpha^3$, $u_2 = \alpha^9$, $u_3 = \alpha^{12}$. Therefore

$$\mathbf{c} = \mathbf{r}_0 - (\alpha^3\, 0\, \alpha^9\, \alpha^{12}\, 0) = (\alpha^3\, \alpha^6\, \alpha^9\, \alpha^{12}\, 1).$$

We are often interested in channels which admit both errors and erasures. The task of the decoder then is to correct the errors and fill the erasure positions. An $[n,M]$-code C over an alphabet A is said to correct t errors and u erasures if for any received vector $\mathbf{r}$ having $l \le u$ erasures there is a unique codeword $\mathbf{c} \in C$ having $d(\mathbf{c},\mathbf{v}) \le t$ for some $\mathbf{v} \in S$, where S is the set of $|A|^l$ vectors which agree with $\mathbf{r}$ in the $n-l$ non-erasure positions. In other words, if any codeword $\mathbf{c} \in C$ suffers at most t errors and up to u additional erasures, then the transmitted codeword $\mathbf{c}$ is recoverable from this altered vector which is received.

Theorem 7.5. An $[n,M]$-code C with distance $d = 2t+u+1$ can correct t errors and u erasures.

Proof.

Let $\mathbf{r}$ be a received vector having at most t errors and $l \le u$ erasures, and let S be the set of vectors which agree with $\mathbf{r}$ in the $n-l$ non-erasure positions. Suppose $\mathbf{c}_1,\mathbf{c}_2 \in C$, $d(\mathbf{c}_1,\mathbf{s}_1) \le t$ and $d(\mathbf{c}_2,\mathbf{s}_2) \le t$ for some $\mathbf{s}_1,\mathbf{s}_2 \in S$. Then by the triangle inequality,

$$\begin{aligned} t+t+u &\ge d(\mathbf{c}_1,\mathbf{s}_1) + d(\mathbf{c}_2,\mathbf{s}_2) + d(\mathbf{s}_2,\mathbf{s}_1) \\ &\ge d(\mathbf{c}_1,\mathbf{s}_1) + d(\mathbf{c}_2,\mathbf{s}_1) \\ &\ge d(\mathbf{c}_1,\mathbf{c}_2). \end{aligned}$$

Now $d = 2t+u+1$ and $d(\mathbf{c}_1,\mathbf{c}_2) \le 2t+u$ implies $\mathbf{c}_1 = \mathbf{c}_2$. This implies there is a unique codeword $\mathbf{c} = \mathbf{c}_1$ that is within distance t of some vector in S, and $\mathbf{r}$ can be decoded to this codeword. Hence C can correct t errors and u erasures. □

The remainder of this section considers decoding linear codes in the presence of both channel errors and erasures. The more specialized case of BCH codes will be discussed in §7.4.

We first consider binary linear codes here. If C is a binary (n,k)-code having distance $d = 2t+u+1$, then C can correct t errors and u erasures. In the presence of no erasures, C will correct up to $t + \lfloor u/2 \rfloor$ errors. Let $\mathbf{r}$ be a received vector having at most t errors and at most u erasures. Suppose the decoder forms two n-tuples $\mathbf{r}_0$ and $\mathbf{r}_1$, where $\mathbf{r}_i$ is obtained from $\mathbf{r}$ by filling all erasure positions with the symbol i, $i = 0,1$. Then since C is binary, in one of $\mathbf{r}_0$ and $\mathbf{r}_1$ at least half the erasure locations now have correct symbols, and hence at least one of $\mathbf{r}_0$ and $\mathbf{r}_1$

has distance at most $t + \lfloor u/2 \rfloor$ from the transmitted codeword. Any standard error correction technique will now correct at least one of these vectors to the transmitted codeword.

If the standard technique decodes both $\mathbf{r}_0$ and $\mathbf{r}_1$ to codewords, and these codewords are the same, then this is the transmitted codeword. If they are different, then that one (and there will be only one) requiring at most t changes to non-erasure positions is the desired codeword. The following examples illustrate these ideas.

Example 5.

Consider the binary (15,7)-code C of example 17 in Chapter 5. Since C has distance 5 it can correct 1 error and 2 erasures, or 2 errors in the absence of erasures. Suppose that

$$\mathbf{c} = (10011\ 01011\ 11000)$$

is a transmitted codeword and that

$$\mathbf{r} = (10\_11\ 00011\ 1100\_)$$

is received. Let

$$\mathbf{r}_0 = (10011\ 00011\ 11000) \quad \text{and} \quad \mathbf{r}_1 = (10111\ 00011\ 11001).$$

Now the standard error-trapping technique of Chapter 5 decodes $\mathbf{r}_0$ to $\mathbf{c}$, and $\mathbf{c}$ differs from $\mathbf{r}$ in only 1 non-erasure position. Hence, assuming at most 1 error (in addition to the 2 erasures) occurred during transmission, $\mathbf{r}$ is correctly decoded to $\mathbf{c}$. The error-trapping technique is unable to decode $\mathbf{r}_1$.

Example 6.

Consider the same code as in the previous example. Suppose the decoder receives

$$\mathbf{r} = (00001\ 00\_\ \_0\ 00000).$$

Then

$$\mathbf{r}_0 = (00001\ 00000\ 00000) \quad \text{and} \quad \mathbf{r}_1 = (00001\ 00110\ 00000).$$

Using error-trapping as the decoding procedure, the decoder decodes $\mathbf{r}_0$ and $\mathbf{r}_1$ to the codewords

$$\mathbf{c}_0 = (00000\ 00000\ 00000) \quad \text{and} \quad \mathbf{c}_1 = (10001\ 01110\ 00000)$$

respectively. Since $\mathbf{c}_0$ differs from $\mathbf{r}$ in 1 non-erasure position, and $\mathbf{c}_1$ differs from $\mathbf{r}$ in 2 non-erasure positions, $\mathbf{r}$ is decoded to $\mathbf{c}_0$.

Now consider linear codes over $GF(q)$. Suppose C is an (n,k)-code with distance $d = 2t + u + 1$, and $\mathbf{r}$ is a received vector with $l \le u$ erasures and at most t additional errors. Let S be the set of q^l n-tuples which agree with $\mathbf{r}$ in the $n - l$ non-erasure positions. The following procedure could be employed to decode $\mathbf{r}$.

Decoding with Erasures for Linear Codes.

(1) Select a vector $\mathbf{s} \in S$.

(2) Determine the unique codeword $\mathbf{c}$ such that $d(\mathbf{c},\mathbf{s}) \le t + \lfloor u/2 \rfloor$, if such a codeword exists. If no such codeword exists, select another $\mathbf{s} \in S$ and repeat step (2).

(3) If $\mathbf{c}$ differs from $\mathbf{s}$ in at most t non-erasure positions, decode $\mathbf{r}$ to $\mathbf{c}$. Otherwise, select another $\mathbf{s} \in S$ and go to (2).

Note that step (1) corresponds to filling values in the erasure positions. This algorithm is practical only for q^u relatively small. For example, if $q = 4$ and $u = 2$, then S contains at most 16 elements.

7.4 BCH Decoding with Erasures

We now consider a procedure for correcting both errors and erasures in BCH codes. The method is similar to the method for error correction of BCH codes given in Chapter 6. In place of the error locator polynomial $\sigma(z)$ we now have an *error-erasure locator polynomial* $\hat{\sigma}(z)$. Suppose there are l_1 errors and l_2 erasures; noting Theorem 7.5, we assume

$$\delta \ge 2l_1 + l_2 + 1 .$$

Then $\hat{\sigma}(z) = \sigma(z)\lambda(z)$ has degree $l = l_1 + l_2$ where

$$\sigma(z) = \prod_{j=0}^{l_1-1} (1 - u_j z)$$

is an error locator polynomial as before and

$$\lambda(z) = \prod_{j=0}^{l_2-1}(1 - u_{l_1+j}z)$$

is the *erasure locator polynomial*. The reciprocals of the roots of $\sigma(z)$ and $\lambda(z)$ determine the error locations and erasure locations, respectively. Letting

$$w(z) = \sum_{i=0}^{l-1} y_i u_i \prod_{\substack{j=0 \\ j\neq i}}^{l-1}(1 - u_j z) \quad \text{and} \quad S(z) = \sum_{i=0}^{\delta-2} S_i z^i$$

as in §6.4, then as before,

$$w(z) \equiv \hat{\sigma}(z)S(z) \equiv \sigma(z)\lambda(z)S(z) \pmod{z^{\delta-1}}.$$

Note $deg\ w(z) = l-1$, $deg\ S(z) \le \delta-2$ and $\lambda(z)S(z)$ has degree at most $l_2+\delta-2$. Let

$$\lambda(z)S(z) = T_1(z) + z^{l_2}T_2(z)$$

where $deg\ T_1(z) \le l_2 - 1$. Then

$$w(z) - \sigma(z)T_1(z) \equiv \sigma(z)z^{l_2}T_2(z) \pmod{z^{\delta-1}}$$

and hence

$$\hat{w}(z) \equiv \sigma(z)T_2(z) \pmod{z^{\delta-l_2-1}} \quad \text{where} \quad \hat{w}(z) = \frac{w(z) - \sigma(z)T_1(z)}{z^{l_2}}$$

and $deg\ \hat{w}(z) \le l_1 - 1$. Now suppose we can find non-zero polynomials $w'(z)$ and $\sigma'(z)$ such that $deg\ w'(z) \le l_1-1$, $deg\ \sigma'(z) \le l_1$ and

$$w'(z) \equiv \sigma'(z)T_2(z) \pmod{z^{\delta-l_2-1}}.$$

Then

$$\hat{w}(z)\sigma'(z) \equiv w'(z)\sigma(z) \pmod{z^{\delta-l_2-1}}$$

and both $\hat{w}(z)\sigma'(z)$ and $w'(z)\sigma(z)$ have degree at most $2l_1 - 1 \le \delta - l_2 - 2$. It follows that

$$\hat{w}(z)\sigma'(z) = w'(z)\sigma(z).$$

Now analogously to §6.4, we can prove $\gcd(\hat{w}(z), \sigma(z)) = 1$ (exercise 5), and that $w'(z)$ and $\sigma'(z)$ are F-multiples of $\hat{w}(z)$, $\sigma(z)$ respectively. It follows that the reciprocals of the roots of $\sigma'(z)$ are the error locators. Knowing these locations, and the original erasure positions, the received vector can be corrected to a codeword by solving a linear system of equations. To find $w'(z)$ and $\sigma'(z)$, we apply the extended Euclidean algorithm to $z^{\delta-l_2-1}$ and $(T_2(z) \bmod z^{\delta-l_2-1})$, continuing until a remainder polynomial $r_i(z)$ is obtained with $deg\, r_i < (\delta - l_2 - 1)/2$. As in §6.4, the existence of such a remainder is guaranteed, here under the assumption that $\delta \geq 2l_1 + l_2 + 1$ (exercise 6). A few examples should help to clarify the ideas presented here.

Example 7.

Consider a BCH code of block length $n = 6$ over $GF(7)$ and designed distance $\delta = 5$ generated by

$$g(x) = (x-\beta)(x-\beta^2)(x-\beta^3)(x-\beta^4)$$

where β is a primitive n^{th} root of unity. Then $g(x)$ generates a (6,2)-RS code C and taking $\beta = 3$,

$$g(x) = (x-3)(x-2)(x-6)(x-4).$$

C can correct up to t errors and u erasures provided

$$\delta \geq 2t + u + 1.$$

Hence C can correct t errors and u erasures for $(t,u) = (0,4)$, (1,2) and (2,0). Suppose the codeword $\mathbf{c} = (4\ 3\ 0\ 5\ 6\ 2)$ is transmitted, and

$$\mathbf{r} = (3\ 3\ \_\ 5\ \_\ 2)$$

is received (i.e. 1 error and 2 erasures occur). Filling erasure positions with 0's yields

$$\mathbf{r}_0 = (3\ 3\ 0\ 5\ 0\ 2) \quad \text{and} \quad r_0(x) = 3 + 3x + 5x^3 + 2x^5.$$

Computing $S_i = r_0(\beta^{i+1}),\ 0 \leq i \leq \delta - 2$ gives

$$S_0 = r_0(3) = (3\ 3\ 0\ 5\ 0\ 2)\cdot(1\ 3\ 2\ 6\ 4\ 5) = 3+2+0+2+0+3 = 3$$

$$S_1 = r_0(2) = (3\ 3\ 0\ 5\ 0\ 2)\cdot(1\ 2\ 4\ 1\ 2\ 4) = 3+6+0+5+0+1 = 1$$

$$S_2 = r_0(6) = (3\ 3\ 0\ 5\ 0\ 2)\cdot(1\ 6\ 1\ 6\ 1\ 6) = 3-3+0-5+0-2 = 0$$

$$S_3 = r_0(4) = (3\ 3\ 0\ 5\ 0\ 2)\cdot(1\ 4\ 2\ 1\ 4\ 2) = 3+5+0+5+0+4 = 3$$

and

$$S(z) = \sum_{i=0}^{\delta-2} S_i z^i = 3 + z + 3z^3.$$

Since the erasures are in locations 2 and 4, the erasure locator polynomial is

$$\lambda(z) = (1-\beta^2 z)(1-\beta^4 z) = (1-2z)(1-4z) = 1 + z + z^2.$$

The error locator polynomial $\sigma(z)$ is not known. Compute

$$\lambda(z)S(z) = 3 + 4z + 4z^2 + 4z^3 + 3z^4 + 3z^5 = T_1(z) + z^{l_2} T_2(z).$$

Now $l_2 = 2$ (number of erasures) gives

$$T_2(z) = 4 + 4z + 3z^2 + 3z^3$$

and

$$\hat{w}(z) \equiv \sigma(z)T_2(z) \equiv \sigma(z)(4+4z) \pmod{z^2}.$$

Applying the extended Euclidean algorithm to $a(z) = z^2$ and $b(z) = 4 + 4z$ we have:

s_i	t_i	r_i	$deg(r_i)$
1	0	z^2	2
0	1	$4z+4$	1
1	$5z+2$	1	0

This yields $\sigma'(z) = 5z + 2$ as a polynomial whose roots are those of $\sigma(z)$. Now $\sigma'(z)$ has root $z = 1 = \beta^0$, implying the error locator $(\beta^0)^{-1} = \beta^0$, giving location number 0. Hence we know that positions 0, 2 and 4 of $\mathbf{r}_0$ are in error. Let

$$e(z) = y_0 + y_2 z^2 + y_4 z^4.$$

Then using $S_i = e(\beta^{i+1}) = r_0(\beta^{i+1})$,

$$S_0 = e(3) = y_0 + 2y_2 + 4y_4 = 3$$

$$S_1 = e(2) = y_0 + 4y_2 + 2y_4 = 1$$

$$S_2 = e(6) = y_0 + y_2 + y_4 = 0$$

and the linear system over $GF(7)$

$$\begin{bmatrix}1 & 2 & 4\\1 & 4 & 2\\1 & 1 & 1\end{bmatrix}\begin{bmatrix}y_0\\y_1\\y_2\end{bmatrix}=\begin{bmatrix}3\\1\\0\end{bmatrix}.$$

This system has solution $(y_0,y_2,y_4) = (6,0,1)$. Hence we correct **r** to

$$\mathbf{r}_0-\mathbf{e} = (3\,3\,0\,5\,0\,2)-(6\,0\,0\,0\,1\,0) = (4\,3\,0\,5\,6\,2).$$

Example 8.

Using the code from the previous example, suppose the codeword $\mathbf{c} = (4\,3\,0\,5\,6\,2)$ is again transmitted and now $\mathbf{r} = (4\,3\,1\,5\,\_\,2)$ is received. Let

$$\mathbf{r}_0 = (4\,3\,1\,5\,0\,2).$$

The erasure locator polynomial is

$$\lambda(z) = (1-\beta^4 z) = (1-4z).$$

We now evaluate the syndromes $S_i = r_0(\beta^{i+1}),\ 0 \le i \le \delta-2$.

$$S_0 = (4\,3\,1\,5\,0\,2)\cdot(1\,3\,2\,6\,4\,5) = 6$$

$$S_1 = (4\,3\,1\,5\,0\,2)\cdot(1\,2\,4\,1\,2\,4) = 6$$

$$S_2 = (4\,3\,1\,5\,0\,2)\cdot(1\,6\,1\,6\,1\,6) = 2$$

$$S_3 = (4\,3\,1\,5\,0\,2)\cdot(1\,4\,2\,1\,4\,2) = 6.$$

Hence

$$S(z) = 6 + 6z + 2z^2 + 6z^3.$$

Now

$$\lambda(z)S(z) = 6 + 3z + 6z^2 + 5z^3 + 4z^4 = T_1(z) + z^{l_2}T_2(z).$$

Since the number of erasures is $l_2 = 1$,

$$T_2(z) = \frac{\lambda(z)S(z)-6}{z} = 3 + 6z + 5z^2 + 4z^3.$$

We now find $w'(z)$ and $\sigma'(z)$ such that

$$w'(z) \equiv \sigma'(z)T_2(z) \pmod{z^3}$$

via the extended Euclidean algorithm:

s_i	t_i	r_i	$deg(r_i)$
1	0	z^3	3
0	1	$5z^2+6z+3$	2
1	$4z+5$	1	0

Hence $\sigma'(z) = 4z + 5$ yields $z = 4$ as a root, implying the error locator $4^{-1} = 2 = \beta^2$ and error location 2. We know now that

$$e(z) = y_2z^2 + y_4z^4.$$

To determine the error magnitudes y_2 and y_4, we obtain the equations

$$S_0 = e(3) = 2y_2 + 4y_4 = 6$$

$$S_1 = e(2) = 4y_2 + 2y_4 = 6.$$

This yields the system

$$\begin{bmatrix} 2 & 4 \\ 4 & 2 \end{bmatrix} \begin{bmatrix} y_2 \\ y_4 \end{bmatrix} = \begin{bmatrix} 6 \\ 6 \end{bmatrix}$$

with solution $(y_2, y_4) = (1,1)$. The vector **r** is corrected to

$$\mathbf{r}_0 - \mathbf{e} = (4\,3\,1\,5\,0\,2) - (0\,0\,1\,0\,1\,0) = (4\,3\,0\,5\,6\,2).$$

Examples 7 and 8 above, and example 16 of §6.4 illustrate a method for determining error magnitudes once the error positions are known. We now outline the general technique. Let C be a BCH code over F of block length n and designed distance δ, with generator polynomial $g(x)$ having β^i, $1 \le i \le \delta - 1$ among its roots, where $\beta \in F$ is a primitive n^{th} root of unity. Suppose $\mathbf{c} \in C$ is transmitted, and $\mathbf{r}$ is received. Let $\mathbf{r}_0$ be the received vector with erasures (if any) filled by 0's, and define $\mathbf{e}$ by the relation

$$\mathbf{r}_0 = \mathbf{c} + \mathbf{e}.$$

The error positions in **r** are determined as discussed earlier and these, together with the erasure positions (which are now treated as error positions in $\mathbf{r}_0$) specify $l \le \delta - 1$ location numbers $i_1, i_2, \ldots, i_l$. Let the corresponding error magnitudes in **e** be y_{i_j}, $1 \le j \le l$ so that

$$e(z) = y_{i_1} z^{i_1} + \cdots + y_{i_l} z^{i_l} = \sum_{j=1}^{l} y_{i_j} z^{i_j}.$$

Now as we have seen before, $e(\beta^i) = r_0(\beta^i)$, $1 \le i \le l$, and hence the set of equations

$$\sum_{j=1}^{l} y_{i_j} (\beta^{i_j})^i = r_0(\beta^i), \quad 1 \le i \le l$$

defines the linear system

$$\begin{bmatrix} \beta^{i_1} & \beta^{i_2} & \cdots & \beta^{i_l} \\ (\beta^{i_1})^2 & (\beta^{i_2})^2 & \cdots & (\beta^{i_l})^2 \\ \vdots & & & \vdots \\ (\beta^{i_1})^l & (\beta^{i_2})^l & \cdots & (\beta^{i_l})^l \end{bmatrix} \begin{bmatrix} y_{i_1} \\ y_{i_2} \\ \vdots \\ y_{i_l} \end{bmatrix} = \begin{bmatrix} r_0(\beta) \\ r_0(\beta^2) \\ \vdots \\ r_0(\beta^l) \end{bmatrix}.$$

Now the coefficient matrix on the left is a Vandermonde matrix, and since the β^{i_j}, $1 \le j \le l$ are distinct and non-zero, this matrix has non-zero determinant (cf. example 13 of Chapter 3, or the proof of Theorem 6.2). The unique solution $(y_{i_1} \cdots y_{i_l})$ is obtained by solving the system, yielding **e** as desired.

7.5 Interleaving

An important technique for increasing the burst error correcting capability of a code is *(codeword) interleaving*. The technique is perhaps most easily explained by example. Consider the binary single error correcting (7,4)-code C with generator matrix

$$G = \begin{bmatrix} 1 & 0 & 0 & 0 & 1 & 1 & 1 \\ 0 & 1 & 0 & 0 & 1 & 1 & 0 \\ 0 & 0 & 1 & 0 & 0 & 1 & 1 \\ 0 & 0 & 0 & 1 & 1 & 0 & 1 \end{bmatrix}.$$

Consider the codewords

$$\mathbf{c}_1 = (1100\,001), \quad \mathbf{c}_2 = (0011\,110), \quad \mathbf{c}_3 = (0111\,000).$$

Suppose we transmit $\mathbf{c}_1$, $\mathbf{c}_2$ and $\mathbf{c}_3$ in this order. Each codeword can correct a single error but none of them can correct even a burst of length 2. The situation can be improved by sending the first bit of $\mathbf{c}_1$ followed by the first bit of $\mathbf{c}_2$ and then the first bit of $\mathbf{c}_3$. We would then send the second bit of $\mathbf{c}_1$ and so on. If we think of the codewords arranged in a 3×7 matrix

$$\begin{matrix} 1 & 1 & 0 & 0 & 0 & 0 & 1 \\ 0 & 0 & 1 & 1 & 1 & 1 & 0 \\ 0 & 1 & 1 & 1 & 0 & 0 & 0 \end{matrix}.$$

then we would transmit the codewords by columns of this matrix. The transmitted sequence would then be

$$100101011011010010100$$

where the three codewords are now *interleaved*. It is easily seen that any burst error of length at most 3 will result in at most 1 error in each of the original codewords. Since these single errors are correctable, the burst error is correctable. Essentially we have taken a (7,4)-code which corrects bursts of length 1 and formed a (21,12)-code which corrects bursts of length 3.

In general, let C be an (n,k)-code over F which corrects any burst of length b. Form a new code C^* as follows. For every collection of t codewords $(a_{i1}\,a_{i2}\,a_{i3} \dots a_{in})$, $1 \le i \le t$, of C form a matrix

$$T = \begin{bmatrix} a_{11} & a_{12} & \cdots & a_{1n} \\ a_{21} & a_{22} & \cdots & a_{2n} \\ \cdot & & & \cdot \\ \cdot & & & \cdot \\ \cdot & & & \cdot \\ a_{t1} & a_{t2} & \cdots & a_{tn} \end{bmatrix},$$

and using the columns of this matrix, form the new codeword

$$(a_{11}\,a_{21} \dots a_{t1}\,a_{12}\,a_{22} \dots a_{t2} \dots a_{1n}\,a_{2n} \dots a_{tn})$$

of C^*. Any burst error of length at most bt will give rise to a burst error of length at most b in any row of T. Since bursts of length b are correctable in the original code, bursts of length bt are correctable in the new codewords. It is easily proven that if we form T in all possible ways the resulting set of codewords is an (nt,kt)-code over F which corrects bursts of length bt. We say that the code is *interleaved to depth t*.

It is interesting to note that applying the preceding construction to a cyclic code results in a cyclic code.

Theorem 7.6. Let C be a cyclic (n,k)-code over $F = GF(q)$ with generator polynomial $g(x)$, capable of correcting bursts of length b. Then $g(x^t)$ is the generator polynomial for a cyclic (nt,kt)-code over F which can correct bursts of length bt.

Proof.

Interleave C to a depth of t to obtain a new code C^*. A codeword in C^* is represented by the $t \times n$ matrix

$$T = \begin{bmatrix} a_0 & a_t & \cdots & a_{(n-1)t} \\ a_1 & a_{t+1} & \cdots & a_{(n-1)t+1} \\ \cdot & \cdot & & \cdot \\ \cdot & \cdot & & \cdot \\ \cdot & \cdot & & \cdot \\ a_{t-1} & a_{2t-1} & \cdots & a_{nt-1} \end{bmatrix},$$

where the rows of T are codewords in C and $a(x) = \sum_{i=0}^{nt-1} a_i x^i$ is the codeword in C^*. C^* has block length nt, has q^{kt} codewords, and by the discussion above, can correct bursts of length bt. It remains to prove that C^* is generated by $g(x^t)$. But note that $a(x)$ may be represented as

$$a(x) = \sum_{i=0}^{t-1} A_i(x^t) x^i$$

where

$$A_i(x^t) = a_i + a_{t+i}x^t + a_{2t+i}x^{2t} + \cdots + a_{(n-1)t+i}x^{(n-1)t}$$

and $A_i(x)$ is a codeword in C. Now since $g(x)$ divides $A_i(x)$, $g(x^t)$ divides $A_i(x^t)$ for $0 \le i \le t-1$. Therefore $g(x^t)$ divides $a(x)$ and hence divides every codeword in C^*. Furthermore, since $g(x)$ divides $x^n - 1$, $g(x^t)$ divides $x^{nt} - 1$ and $g(x^t)$ generates a linear code of dimension kt. It follows that $g(x^t)$ is the generator polynomial for C^*. (See also exercise 11, Chapter 6.) □

We now consider a variation of interleaving, known as *cross-interleaving*, which interleaves several codes. Cross-interleaving is an important technique in practice. As before, we find it most convenient to proceed by way of example.

Consider again the binary single-error correcting (7,4)-code C_1 with generator matrix

$$G_1 = \begin{bmatrix} 1&0&0&0&1&1&1 \\ 0&1&0&0&1&1&0 \\ 0&0&1&0&0&1&1 \\ 0&0&0&1&1&0&1 \end{bmatrix}$$

and now also the binary single-error correcting (6,3)-code C_2 with generator matrix

$$G_2 = \begin{bmatrix} 1&0&0&1&1&1 \\ 0&1&0&1&1&0 \\ 0&0&1&0&1&1 \end{bmatrix}.$$

Suppose that we interleave C_1 to a depth of 3, but instead of using each column of the 3 by 7 matrix as part of the new codeword as before, we now take each column as the 3 information symbols for a codeword in C_2. For example, the C_1 codewords

$$\mathbf{c}_1 = (1\,1\,1\,1\,1\,1\,1), \quad \mathbf{c}_2 = (0\,1\,1\,1\,0\,0\,0), \quad \mathbf{c}_3 = (0\,1\,0\,0\,1\,1\,0)$$

would be interleaved using the matrix

$$T = \begin{bmatrix} 1&1&1&1&1&1&1 \\ 0&1&1&1&0&0&0 \\ 0&1&0&0&1&1&0 \end{bmatrix}.$$

The first column is encoded to 1 0 0 1 1 1 using G_2, the second column to 1 1 1 0 1 0 and so on. The seven columns produce the seven C_2 codewords

$$\begin{array}{c} 100\ 111 \\ 111\ 010 \\ 110\ 001 \\ 110\ 001 \\ 101\ 100 \\ 101\ 100 \\ 100\ 111 \end{array} \tag{1}$$

We could now interleave these codewords to any depth desired. Suppose we interleave to a depth of 2. The resulting symbol string will be

$$1101011011101111000000111100111110000. \qquad (2)$$

Since there is no codeword to pair with the seventh vector, we will interleave it with the first word from the next set of interleaved codewords of C_1.

As we have seen, a code with distance d can correct $d-1$ erasures, but can correct only $(d-1)/2$ errors. Clearly erasures are easier to correct than errors. A decoding strategy for a cross-interleaved code can take advantage of this fact by using the error detection capabilities of the outer code (along with, possibly, some error correction) together with the erasure-correcting capabilities of the inner code. The interleaving of the inner codewords has the effect of dispersing errors (erasures) among *inner* codewords.

Returning to our example, using the error correction capabilities of code C_2 alone, any burst of length 2 can be corrected in the resulting cross-interleaved string. Now suppose C_2 was used exclusively for error detection, and C_1 was used to correct erasures only. Having distance 3, C_2 can detect 2 errors, and C_1 can correct 2 erasures. When a received bit sequence (such as (2)) is de-interleaved to produce (supposed) C_2 codewords (as in (1)), the code is used to detect errors in a C_2 codeword; if errors are detected, all information symbols from that codeword are marked as erasures and sent on to be de-interleaved to get C_1 codewords. In our example, if at most 2 successive sets of 3 information symbols (corresponding to a burst error of length 12) are flagged as erasures and passed on – say the information symbols

100
111

from the first 2 codewords in (1) – then de-interleaving results in at most 2 erasures in each resulting C_1 codeword:

$$c_1 = (\_\,\_\,1\,1\,1\,1\,1), \quad c_2 = (\_\,\_\,1\,1\,0\,0\,0), \quad c_3 = (\_\,\_\,0\,0\,1\,1\,0).$$

These erasures can then be corrected by C_1 (providing there are no additional errors).

Notice that the outer interleaving (i.e. interleaving of C_2 codewords) does not affect the correction capabilities of C_1 directly, but disperses a burst error over C_2 codewords. In the case that C_2 is used for correcting a small number of errors (even a single error) and detecting further errors simultaneously, outer interleaving improves the correction capability of C_2 itself for short random bursts.

In actual implementations, interleaving may be done using a slightly different technique, known as *delayed interleaving*. In this case, to interleave an (n,k)-code C to a depth n, rather than interleaving fixed blocks of n codewords as above, the codewords are interleaved in a continuous sequence. If $\mathbf{c}_i = (c_{i,1}\, c_{i,2} \ldots c_{i,n})$ is the ith codeword of the sequence to be interleaved, $i > n$, then for the purposes of interleaving consider the array

$$\begin{array}{cccccccccc}
c_{i-(n-1),1} & \cdots & c_{i-2,1} & c_{i-1,1} & c_{i,1} & c_{i+1,1} & & & & \\
& c_{i-(n-1),2} & \cdots & c_{i-2,2} & c_{i-1,2} & c_{i,2} & c_{i+1,2} & & & \\
& & & \cdot & \cdot & & & & & \\
& & & \cdot & c_{i-1,3} & c_{i,3} & c_{i+1,3} & & & \\
& & & \cdot & \cdot & & & & & \\
& & & c_{i-(n-1),n-1} & c_{i-(n-2),n-1} & & \cdots & \cdots & \cdots & \\
& & & & c_{i-(n-1),n} & \cdots & & c_{i-1,n} & c_{i,n} & c_{i+1,n}
\end{array}$$

The interleaved string of symbols is now derived from the columns of this array. Since the codewords from C are distributed down diagonals of this array, it is clear that some initialization must take place to guarantee complete columns at the beginning and end of the process; for example, the columns could be padded out with 0's. Once initialization is complete, each successive codeword of the codeword sequence to be interleaved completes a column of the array, and hence completes an interleaved codeword or *frame*. Without the delay technique, a column of the interleave array would be completed only after every n codewords, and at that time n columns would be completed at once; the continuous stream obtained by delayed interleaving is preferable.

Note that if C is an (n,k)-code capable of correcting a single error (per codeword), then C must be interleaved to a depth of at least n to guarantee that a burst error of length n is correctable. In the array above, this translates to having no column contain more than one symbol from any one codeword of C.

Using delayed interleaving as governed by the array above, consecutive symbols within a codeword end up in consecutive frames (i.e. one frame apart) after interleaving. We call this a *1-frame delay interleave*. To generalize this to a *d-frame delay interleave*, so that consecutive symbols end up d frames apart for any fixed $d \geq 1$, rather than placing symbol $c_{i,j+1}$ one column over and one row down from $c_{i,j}$ as in the array above, $c_{i,j+1}$ is placed d columns over and one row down from $c_{i,j}$.

7.6 Error Correction and Digital Audio Recording

Digital audio has been heralded by many as the greatest advancement in sound technology since Thomas Edison's invention of the phonograph in 1877. The role played by error-correcting codes in this area is by no means insignificant. Due to the vast amounts of data required to represent audio signals digitally, and the resulting requirement of a very high density storage medium for digital audio recordings, powerful error correction and suppression capabilities are essential to make this technology commercially viable. In this section we will give an overview of how errors are handled in one of the most popular digital storage mediums to date, that of the Compact Disc digital audio system developed jointly by N.V. Philips of The Netherlands and Sony Corporation of Japan. We begin with a brief description of the system.

The storage medium used is a *compact disc* (CD), a flat circular disc resembling a conventional phonograph record but less than 5 inches in diameter, aluminized (for reflectivity, as we shall see is necessary) and coated with a clear protective plastic. Rather than representing an audio signal as a continuous waveform, for digital recordings the signal is sampled at fixed time intervals, quantized and stored as a sequence of binary numbers. The method of coding sound for storage and playback is called *(linear) pulse code modulation* (PCM). At a given instance in time the sound wave is sampled, and the amplitude is determined and assigned a discrete value from 1 to $2^{16}-1$. This value is given as a binary 16-tuple. Actually two samples, one for the left channel and one for the right, are taken. These samples are taken at a rate of 44,100 per second (44.1 kHz). Each binary 16-tuple is taken to represent two field elements from $GF(2^8)$, and hence each sample produces 4 $GF(2^8)$ symbols. On playback, the compact disc player will have to process $(44{,}100)(32) = 1{,}411{,}200$ bits of audio data per second. As will be seen shortly, for various reasons the number of bits actually processed per second is substantially higher than this.

For purposes of error correction, information is grouped into segments called *frames*, with each frame holding 24 data symbols. The code used for error correction is a *Cross-Interleaved Reed-Solomon Code* (CIRC), obtained by cross-interleaving two Reed-Solomon codes. The 24 symbols from $GF(2^8)$ from 6 samples are used as information symbols in a (28,24)-RS code C_1 over $GF(2^8)$. The code is now interleaved to a depth of 28 using a 4-frame delay interleave. The resulting columns of 28 symbols are used as information symbols in a (32,28)-RS code C_2 over $GF(2^8)$, with 4 additional parity check symbols determined by each column. (We note that these codes are actually *shortened Reed-Solomon codes*. Shortened codes are defined and discussed in exercises 27-38). The symbols in a frame, which now corresponds to a C_2 codeword, are then regrouped to separate the odd and even-numbered symbols of that frame into distinct frames, with the symbols in odd-numbered symbol positions of one frame grouped with symbols in even-numbered positions from the next frame in time to form a new frame.

At this stage the frame consists of 32 8-bit symbols (24 audio data symbols and 8 parity symbols). One more 8-bit symbol is added which holds "control and display" information, including information for the disc directory and unused bits (possibly for future use). In order to complete the bit description of a frame we require some knowledge of how the information is stored and read from the disc itself.

The disc contains a spiral track in which depressions or *pits* have been made. Flat areas between pits are called *lands*. The audio information is encoded by the lengths of these pits and that of the lands separating them. A transition from a land to a pit or vice versa (i.e. a pit wall) is read as a 1, and distances along a land or pit are read as strings of 0's. For example, the following three pits might represent the associated 0,1 string.

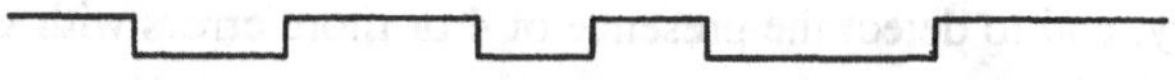

000 1000 10000 100 100 100000 1000000

These transitions can be determined by changes in the intensity of reflected light from a laser beam tracking the sequence of pits and lands. For technical reasons, and to maximize the amount of data that is placed on the disc, it is required that the minimum distance between transitions (***minimum run-length***) be 3 channel bits, and the maximum distance between transitions (***maximum run-length***) be 11 channel bits. This implies that the data recorded on the disc be a binary string having between 2 and 10 0's between successive 1's. The bits in the frame as currently specified will not in general have these properties, so it is necessary to modulate the data further. To this end each 8-bit symbol is assigned a unique 14-bit sequence with the desired properties. This is called ***eight-to-fourteen modulation (EFM)***. Each of these 14-bit strings receives an additional 3 bits to merge it with the next 14-bit string. Hence each 8-bit sequences is mapped to a 17-bit sequence. The conversion is done by a table lookup.

A frame is completed by adding a 24-bit synchronization pattern and 3 additional merging bits. The total number of bits for a frame is then

$$(32)(17) + (17) + (27) = 588.$$

These are called the ***channel bits***, and are the bits actually recorded on the disc. Now since a frame represents 6 samplings, and sampling is at 44.1 kilohertz, $(44100/6)(588) = 4{,}321{,}800$ channel bits are required for each second of play. Most CDs can hold about 74 minutes of music, which translates to $(74)(60)(4321800) = 19{,}188{,}792{,}000$ channel bits. Of course, the actual number of *audio bits* is about 1/3 of this number. Due to the effective use of EFM, these channel bits require only about 2 billion pits on the disc.

Since each pit is only about 0.5 micrometers wide, and the length of a pit or the land between pits varies between 0.9 and 3.3 micrometers (depending on the number of 0's the pit represents), fingerprints, dust, dirt and surface abrasions and irregularities can lead to numerous read errors. On playback, after synchronization bits and merging bits are removed, and the 14-bit segments are demodulated, the 32 symbols of a frame are passed in parallel to be de-interleaved. (Recall that the odd and even-numbered symbols of alternate frames were interleaved when stored on the disc.) The resulting frames are then processed by the decoder for C_2. C_2 has distance 5 and is hence capable of correcting 2 errors. However, the decoder is designed to correct only single errors. This makes it possible to simultaneously detect the presence of 2 or 3 errors with certainty, and to detect the presence of 4 or more errors with very high probability. The ability of C_2 to correct any single error and simultaneously detect up to 3 errors follows from the fact that it has distance 5. The probability that 4 or more errors will go undetected can be determined as follows. The number of vectors in spheres of radius 1 about codewords is $(q^k)(1+n(q-1))$, where here $n=32$, $k=28$ and $q=2^8$. There are in total q^n n-tuples over $GF(q)$. Errors moving one codeword into the sphere of radius 1 about another codeword will go undetected. Hence the probability of errors going undetected is approximately

$$\frac{q^k(1+n(q-1))}{q^n} = \frac{32q-31}{q^4} \sim 2^{-19}.$$

Notice that if we would use the full error correcting capability of the code the probability of undetected errors becomes approximately

$$\frac{q^k(1+n(q-1)+\binom{n}{2}(q-1)^2)}{q^n} = \frac{32q-31+(16)(31)(q-1)^2}{q^4} \sim 2^{-7},$$

which is substantially higher.

If the decoder for C_2 detects the presence of a single error, the error is corrected. If the presence of a multiple error is detected, the decoder flags all 28 information symbols in the frame as erasures, and sends these on to the decoder for C_1. (The parity symbols of C_2 are discarded at this point.) Before reaching the decoder for C_1, the 28 symbols are delay de-interleaved. C_1 also has distance 5 and hence can correct up to 2 errors or 4 erasures per C_1 codeword. The decoder for C_1 is designed to perform the latter. Due to the 4-frame delay interleave, the C_1 decoder can thus correct burst errors that result in the decoder for C_2 marking all symbols in up to 16 consecutive disc frames as erasures. This corresponds to a burst error of about 4000 *data bits* (audio bits plus parity bits) or 3000 audio bits, which translates to a track length of about 2.5 mm on the disc.

We note again that code C_2 is used to both correct short random burst errors itself and to further detect errors and exploit the erasure correcting capabilities of code C_1. The even-odd interleaving (grouping) of symbols from successive C_2 codewords enhances the capability to carry out the former, while the 4-frame delay interleave facilitates the latter.

Even in the event that an erroneous sample value that has been flagged as unreliable by the decoder for C_2 cannot be corrected by the erasure-correction of the decoder for C_1, the error may still be *concealed* if the sample values from the sampling instant immediately before and immediately after the sample in question are reliable. In this case the unreliable sample value can be replaced by the value determined by linear interpolation of these reliable neighbours. To maximize the length of the largest burst-error that can be concealed by linear interpolation, before the parity check bits for a C_1 codeword are determined the sample sets (L_i,R_i), $i = 1, ..., 6$ (corresponding to left and right channel samples at 6 consecutive sampling instances) of the frame are permuted to separate even and odd numbered (i.e. neighbouring) samples as far as possible within that frame, as follows:

$$L_1\, L_3\, L_5\, R_1\, R_3\, R_5\, P\ P\ L_2\, L_4\, L_6\, R_2\, R_4\, R_6$$

The P's are then computed as the parity symbols (2 symbols each) for the C_1 codeword the frame corresponds to. Without going further into details here, we note that this permutation coupled with the 4-frame delay interleave makes possible interpolation to conceal burst errors of over 12,000 data bits, corresponding to a track length of more than 7.5 mm on the disc.

In the unlikely event that the decoder for C_2 fails to detect an error but the decoder for C_1 does detect such an error, then all 24 audio symbols of the word in question are flagged as being unreliable. This highly improbable situation might arise due to uncorrected random errors, even in the absence of long burst errors. With the scheme as described thus far, such a rejection of a single entire frame by the C_1 decoder would make error concealment by interpolation impossible, as at least one neighbour of each sample tagged as unreliable would also be unreliable.

For this reason, before the permutation of symbols within a frame (as described above) is carried out during encoding, the odd-numbered samples (left and right samples numbered 1, 3 and 5) in each sampling frame of 6 consecutive sample sets are separated from the even-numbered samples, with the odd-numbered samples of each frame grouped with the even-numbered samples from 2 frames later in time to form a new frame. Consecutive (i.e.

neighbouring) samples are then 2 frames apart. This makes possible error concealment via interpolation from reliable neighbouring samples even in the event that the decoder for C_1 rejects two entire consecutive frames.

In the case that both error correction and error concealment by interpolation are not possible, errors are concealed by muting the sounds associated with erroneous symbols. The interested reader may find further details in the literature [Hoeve 82], [Vries 82], [Pohlmann 87].

7.7 Exercises

Reed-Solomon codes

1. For each of the following fields F determine all values n and k for which an (n,k)-RS code over F exists. In each case state the error-correcting capabilities of the code.
   (a) $F = GF(2^6)$ (b) $F = GF(2^8)$ (c) $F = GF(2^9)$ (d) $F = GF(3^2)$
   (e) $F = GF(3^3)$ (f) $F = GF(5^2)$ (g) $F = GF(7^2)$ (h) $F = GF(7^3)$
   (i) $F = GF(37)$ (j) $F = GF(79)$ (k) $F = GF(101)$

2. Prove that the dual of an (n,k)-RS code over F is an $(n,n-k)$-RS code over F.

3. (a) Give a generator polynomial for a double-error correcting (10,5)-*RS* code C over $GF(11)$.

   (b) Determine the distance in C.

   (c) Construct a parity check matrix H for C.

Erasures and decoding

4. The decoding technique of §7.4 essentially involves the reduction of the error-erasure locator polynomial to an error locator polynomial, removing the erasure locator polynomial $\lambda(z)$ from consideration and thereby reducing the problem to that of §6.4. Suppose $\lambda(z)$ is in fact also an error-locator polynomial (of degree l_2), i.e. no erasures occur but now $l = l_1 + l_2$ errors occur. Can the technique of §7.4 be applied to remove $\lambda(z)$ from consideration now?

5. Using the notation of §7.4, prove that $\gcd(\hat{w}(z), \sigma(z)) = 1$.

6. Again using the notation of §7.4, and assuming $\delta \geq 2l_1 + l_2 + 1$, prove that in the remainder sequence obtained by applying the extended Euclidean algorithm to $z^{\delta-l_2-1}$ and $T_2(z)$, a remainder polynomial $r_i(z)$ with $deg\ r_i < (\delta-l_2-1)/2$ is guaranteed.

7. Let C be a linear code over F whose distance is d. Suppose a codeword $\mathbf{c}$ is transmitted and the vector $\mathbf{r}$ is received with $l \leq d-1$ erasures, say in the first l positions, and no other errors. Let $\mathbf{r}_0$ be the vector obtained from $\mathbf{r}$ by filling the erasure positions with 0's, and let $\mathbf{r}_0 = \mathbf{c} + (y_1\ y_2 ... y_l 0 ... 0)$. Let H be a parity check matrix for C, with i^{th} column denoted $\mathbf{h}_i$. Prove that the system of linear equations given by $H\mathbf{r}_0^T = \sum_{i=1}^{i=l} y_i \mathbf{h}_i$ has a unique solution, and hence can be used to determine the erasure magnitudes.

8. Suppose C is a (7,4)-cyclic binary code generated by $g(x) = 1+x+x^3$ and C can correct up to l erasures. Determine the largest value of l which is possible.

9. Consider $GF(2^6)$ generated by a root α of $f(x) = 1+x+x^6$. The polynomial $g(x) = (x-\alpha^7)(x-\alpha^{14})(x-\alpha^{21})(x-\alpha^{28}) = x^4+\alpha^{22}x^3+\alpha^{26}x^2+\alpha^{57}x+\alpha^7$ generates a (9,5)–*RS* code having distance 5. Suppose that a codeword **c** is transmitted and only erasures are introduced by the channel. If $\mathbf{r} = (\ \alpha^{57}\ 1\ \alpha^{35}\ \alpha^7\ \alpha^{22}\ \_\ 0\ 0)$ is received, determine **c**.

10. A (7,4)-cyclic binary code C is generated by $g(x) = 1+x+x^3$. Decode $\mathbf{r} = (0\ 1\ \_\ 0\ 0\ \_\ 1)$.

11. Consider the code given in example 8, Chapter 6.

    (a) Determine the erasure and error combinations that the code can correct.

    (b) Decode

    (i) $\mathbf{r}_1 = (0\ 1\ 1\ 0\ \_\ 0\ 1\ 1\ 1\ \_\ 1\ 1\ 0\ 0\ 1)$

    (ii) $\mathbf{r}_2 = (1\ 0\ 1\ 0\ 0\ \_\ 1\ 0\ \_\ 1\ \_\ 0\ 1\ 0\ 1)$

    (iii) $\mathbf{r}_3 = (1\ 0\ \_\ 1\ \_\ 0\ \_\ \ \_\ 1\ 0\ 0\ \_\ 0\ 1\ 1)$.

12. Let $g(x) = \prod_{i=1}^{6}(x-\beta^i)$ generate a (10,4)–*RS* code C where $\beta = 2$ is a primitive element in $GF(11)$. Consider the received vector $\mathbf{r} = (3\ 8\ 6\ \_\ 9\ 9\ \_\ 2\ 2\ 0)$.

    (a) Determine the error and erasure combinations that C can correct.

    (b) Find the symdrome polynomial $s(z)$ for **r**.

    (c) Find the error-erasure locator polynomial $\hat{\sigma}(z)$.

    (d) Find the roots of $\hat{\sigma}(z)$.

    (e) Decode **r**.

13. Consider a (12,8)–*RS* code C generated by $g(x) = \prod_{i=1}^{5}(x-\beta^i)$ where $\beta = 7$ is a primitive element in $GF(13)$. Decode $\mathbf{r} = (a\ 3\ 7\ \_\ a\ 1\ 8\ c\ b\ 7\ \_\ 1)$ where $a = 10$, $b = 11$ and $c = 12$.

Interleaving and decoding strategies

14. Let

$$H = \begin{bmatrix} 1 & 1 & 0 & 1 & 1 & 0 & 0 \\ 0 & 1 & 1 & 1 & 0 & 1 & 0 \\ 1 & 1 & 1 & 0 & 0 & 0 & 1 \end{bmatrix}$$

be a parity-check matrix for a binary (7,4)-code C.

(a) Encode the following information tuples

(0 1 1 0), (1 0 1 0), (1 0 0 0), (1 0 1 1), (0 1 1 1), (0 1 0 0), (0 0 1 1).

(b) What will be the sequence of transmitted symbols using delayed interleaving to a depth of 7? (Fill sequence with 0's where necessary.)

15. Let $G = \begin{bmatrix} 1 & 1 & 0 & 1 & 0 \\ 1 & 0 & 1 & 0 & 1 \end{bmatrix}$ be a generator matrix for a binary (5,2)-single-error-correcting code C. Suppose the following is a sequence of received symbols formed by delayed interleaving codewords in C to a depth of 5 (where 0's have been added accordingly):

1 0 0 0 0 0 0 0 0 0 1 1 1 0 0 1 0 0 0 0 1 1 0 1 1
0 0 0 1 1 0 0 1 1 0 0 0 0 0 0 1 0 0 0 1

From this sequence determine the transmitted sequence.

16.

$$G = \begin{bmatrix} & 1 & 0 & 1 & 0 \\ & 0 & 1 & 1 & 0 \\ I_4 & 1 & 0 & 0 & 1 \\ & 0 & 1 & 0 & 1 \end{bmatrix}$$

is a generator matrix for a binary (8,4)-code C having distance 3. Three codewords were interleaved and transmitted. Determine the error pattern associated with the received sequence

11001 11001 11010 01010 1101

17. Let C_1 be the code given in exercise 16 above and let C_2 be a binary (7,3)-code generated by

$$\begin{bmatrix} & 1\,0\,1\,1 \\ I_3 & 0\,1\,0\,1 \\ & 1\,1\,0\,0 \end{bmatrix}.$$

Suppose 3 codewords from C_1 are cross-interleaved with C_2 and codewords in C_2 are interleaved to a depth of 4. Determine the error pattern associated with the received sequence:

1 0 0 0 0 1 1 1 0 1 1 1 1 1 0 1 0 0 1 0 1 0 0 0 1
1 1 0 0 0 1 0 1 0 1 1 0 1 1 1 1 0 0 1 1 0 0 1 1 0
1 0 0 0 0 0

(Note: Spaces have been introduced into the sequence for ease of reading only.)

18. While concealment of errors by linear interpolation is useful in digital audio applications, why is this technique not useful in the general context of error-correcting codes?

19. Determine the length (in bits) of the longest burst error that is always correctable by the CIRC and error-correction strategy as outlined in §7.6.

Weight Distribution and Enumerators

20. Let C be an (n,k)-code over $GF(q)$ having distance $d = n-k+1$ (C is an *MDS* code). Prove that C contains a codeword of minimum weight where the d nonzero entries are in any preselected d coordinate positions.

21. Prove that the converse of exercise 20 is also true. That is, if C is an (n,k)-code over $GF(q)$ having the property that for any d coordinate positions there is a vector of minimum weight having non-zero entries only in these positions then $d = n-k+1$.

22. For the code C given in exercise 20 prove that the number of codewords of weight $n-k+1$ is $(q-1)\binom{n}{n-k+1}$.

23. Prove that the *MDS* code C of exercise 20 has k distinct nonzero weights $n-k+1,\ldots,n$ and that $C^\perp$ has minimum distance $k+1$.

24. Let C be the code given in exercise 20 and let $(A_0,A_1,\ldots,A_n)$ be its weight distribution. Recall from Chapter 3, exercise 84 the identity

$$\sum_{i=0}^{n-j}\binom{n-i}{j}A_i = q^{k-j}\sum_{i=0}^{j}\binom{n-i}{j-i}A_i', \quad j = 0,1,\ldots,n.$$

By exercise 23 we have that $A_i = 0$, $1 \le i \le n-k$ and $A_i' = 0$, $1 \le i \le k$. Deduce that

(a) $\sum_{i=n-k+1}^{n-j} \binom{n-i}{j} A_i = \binom{n}{j}(q^{k-j}-1), \quad j=0,1,...,k-1.$

(b) $A_{n-k+1} = \binom{n}{k-1}(q-1).$

(c) $A_{n-k+2} = \binom{n}{k-2}((q^2-1)-(n-k+2)(q-1)).$

(d) $A_{n-k+r} = \binom{n}{k-r} \sum_{j=0}^{r-1} (-1)^j \binom{n-k+r}{j}(q^{r-j}-1), \quad r \geq 3.$

(Note: This problem shows how to determine the weight distribution for any Reed-Solomon code since these codes are examples of *MDS* codes.)

25. Determine the weight enumerator for the code given in exercise 3.

26. Determine the weight distribution for the code given in exercise 13.

Shortened Codes

27. Let C be an (n,k)-code over F having generator matrix G in standard form. Let t be a positive integer such that $t \leq k$ and let C' be the set of all codewords having 0's in the first t coordinate positions. Prove that C' is an $(n,k-t)$-code over F. (If we delete the first t coordinates from each vector in C' then the resulting set of vectors forms an $(n-t,k-t)$-code over F which is called *a shortened code*.)

28. If C and C' are as described above and they have distances d and d' respectively,

    (a) prove that $d' \geq d$;

    (b) show, by constructing an example, that it is possible for $d' > d$.

29. Let C be an (n,k)-cyclic code over F. Let C' be an $(n-t,k-t)$-shortened cyclic code obtained from C. Show by example that C' need not be a cyclic code.

30. (a) Construct a (6,4)-*RS* code C over $GF(7)$ using the primitive element 3.

    (b) From the code C constructed in (a) form a (4,2)-shortened cyclic code and determine its distance.

31. Prove that a shortened Reed-Solomon code has the same distance as the original code.

32. Let C be a cyclic code over F of block length n generated by $g(x)$. Let C' be an $(n-t,k-t)$-shortened cyclic code obtained from C. Prove that there exists a polynomial $f(x) \in F[x]$ such that C' is an ideal in $F[x]/(f(x))$.

33. Let C' be an ideal in $F[x]/(f(x))$ for some polynomial $f(x)$ having degree $n' \geq 3$ and let $g(x)$ be a generator for C'. Prove that if C' contains no vectors of weight 2 then C' is a shortened cyclic code.

34. Construct a generator matrix for a binary (13,5)-code which has distance 5.

35. Construct a generator matrix for a binary (13,3)-code which corrects 3 errors.

36. Let C be the (13,6)-code over $GF(3)$ described in example 16, Chapter 6. Let C' be the (11,4)-code obtain from C by shortening. Decode the following received vectors:

    (i) $\mathbf{r}_1 = (1\,0\,1\,1\,0\,1\,1\,0\,2\,2\,0)$

    (ii) $\mathbf{r}_2 = (1\,2\,2\,1\,1\,1\,2\,2\,2\,2\,2)$.

37. Determine the weight distribution for both of the codes C and C' given in exercise 36.

38. Let C be an (n,k)–RS code over F and let C' be a shortened $(n-t,k-t)$-code obtained from C. Can you determine the weight distribution of C'? Explain your reasoning.

Appendix A

REVIEW OF VECTOR SPACES

A vector space V is defined with respect to a field F. The elements of V are called *vectors;* the elements of F are called *scalars*. The vector space has two (binary) operations, which are often written "+" and "·", denoting vector addition (for addition of two vectors) and scalar multiplication (for multiplication of a vector by a scalar) respectively. The operations of addition and multiplication within the field F itself are in general distinct from these vector space operations, although the same symbols are often used for both, and "·" is often simply indicated by juxtaposition of the two operands. As a formal definition, we have the following.

Definition. A *vector space* over a field F is a set of elements V together with two operations "+" and "·" satisfying the following axioms for all $\mathbf{x}, \mathbf{y}, \mathbf{z} \in V$ and all $a, b \in F$:

1. $\mathbf{x}+\mathbf{y} \in V$
2. $\mathbf{x}+\mathbf{y}=\mathbf{y}+\mathbf{x}$
3. $\mathbf{x}+(\mathbf{y}+\mathbf{z}) = (\mathbf{x}+\mathbf{y})+\mathbf{z}$
4. there exists an element $\mathbf{0} \in V$ such that $\mathbf{x}+\mathbf{0}=\mathbf{x}$
5. there exists an element $\bar{\mathbf{x}}$ such that $\mathbf{x}+\bar{\mathbf{x}}=\mathbf{0}$
6. $a \cdot \mathbf{x} \in V$
7. for the scalar $1 \in F$, $1 \cdot \mathbf{x}=\mathbf{x}$
8. $a \cdot (\mathbf{x}+\mathbf{y})=a \cdot \mathbf{x}+a \cdot \mathbf{y}$
9. $(a+b) \cdot \mathbf{x}=a \cdot \mathbf{x}+b \cdot \mathbf{x}$
10. $(ab) \cdot \mathbf{x}=a \cdot (b\mathbf{x})$

The axioms 1-5 specify that $(V,+)$ is an Abelian group under vector addition. Axiom 4 specifies the existence of a *zero vector* or additive identity. Axiom 5 specifies the existence of an additive inverse for each vector; $\bar{\mathbf{x}}$ is often denoted $-\mathbf{x}$. Axiom 6 specifies closure under scalar multiplication. Axioms 8 and 9 determine how the two operations distribute over one another, and axiom 10 specifies that scalar multiplication is associative.

We now consider several examples. The set of ordered n-tuples over a field, for example the field $\mathbf{R}$ of real numbers, together with the standard operations of componentwise addition and multiplication of an n-tuple by a scalar from $\mathbf{R}$, forms a vector space over $\mathbf{R}$ known as *Euclidean n-space* over $\mathbf{R}$, often denoted $\mathbf{R}^n$. The field $\mathbf{R}$ here may be replaced by any field, including a finite field F.

The set $\mathbf{P}_n(x)$ of polynomials in the indeterminate x of degree at most n, with coefficients from $\mathbf{R}$, together with polynomial addition and scalar multiplication defined in the ordinary way, i.e. for $f(x), g(x) \in \mathbf{P}_n(x)$ and $r \in \mathbf{R}$,

$$(f+g)(x) = f(x)+g(x)$$

$$(r\cdot f)(x) = r\cdot f(x)$$

forms a vector space over $\mathbf{R}$. (How does this vector space compare to $\mathbf{R}^n$?)

As a third and final example, the set of $m \times n$ matrices over $\mathbf{R}$, with the operations of matrix addition and matrix multiplication by a scalar, forms a vector space often denoted $\mathbf{R}_{mn}$.

We now review some standard definitions and terminology related to vector spaces. If V is a vector space over a field F and $\mathbf{x} \in V$, then $\mathbf{x}$ is said to be a *linear combination* of the vectors $\mathbf{y}_1, \mathbf{y}_2, \ldots, \mathbf{y}_s \in V$ over F if for some scalars $a_1, a_2, \ldots, a_s \in F$,

$$\mathbf{x} = \sum_{i=1}^{s} a_i \cdot \mathbf{y}_i .$$

A set of elements $\{\mathbf{y}_1, \mathbf{y}_2, \ldots, \mathbf{y}_s\}$ is *linearly independent* (over F) if the only solution to the equation

$$a_1\cdot\mathbf{y}_1 + a_2\cdot\mathbf{y}_2 + \cdots + a_s\cdot\mathbf{y}_s = \mathbf{0}$$

is the trivial solution $a_1 = a_2 = \cdots = a_s = 0$. Otherwise, these elements are said to be *linearly dependent*. The *span* (over F) of a set of vectors is the set of all linear combinations (over F) of the vectors in the set. A set of vectors $\{\mathbf{y}_1, \mathbf{y}_2, \ldots, \mathbf{y}_s\}$ is called a *basis* for V if each vector $\mathbf{x} \in V$ can be uniquely expressed as a linear combination of these vectors, i.e. if for each $\mathbf{x} \in V$, $\mathbf{x} = a_1\cdot\mathbf{y}_1 + \cdots + a_s\cdot\mathbf{y}_s$ for unique scalars $a_1, \ldots, a_s \in F$. It follows from the definition that the vectors in a basis are linearly independent, and furthermore, that any set of linearly independent vectors whose span is the entire space V is a basis for V. All bases for a fixed vector space V over F contain the same number of elements, and the number of elements in a basis is called the *dimension* of V (over F).

Appendix B

THE DIVISION ALGORITHM AND THE EUCLIDEAN ALGORITHM

We summarize here the division algorithm, the Euclidean algorithm and the extended Euclidean algorithm for integers, and the analogous algorithms for polynomials. While our interest lies mainly with the algorithms for polynomials, the algorithms for the integer case are perhaps already familiar to many readers, and in any case, can be easily understood, and generalize directly to the polynomial case. Further details regarding the material presented here may be found in any elementary book on modern algebra, for example [Gilbert 76], or among other references, in [McEliece 87].

The division algorithm for integers. Given positive integers a and $b \neq 0$, their exist unique integers q and r such that

$$a = q \cdot b + r \qquad \text{where } 0 \leq r < b .$$

The *quotient* q and *remainder* r can be easily determined by standard "long division".

The Euclidean algorithm for integers. Given positive integers a and $b \neq 0$, their greatest common divisor gcd(a,b) can be found through repeated use of the division algorithm, as the last nonzero remainder r_k in the sequence

$$\begin{aligned}
a &= q_2 \cdot b + r_2, & 0 < r_2 < b \\
b &= q_3 \cdot r_2 + r_3, & 0 < r_3 < r_2 \\
r_2 &= q_4 \cdot r_3 + r_4, & 0 < r_4 < r_3 \\
&\cdots & \\
r_{k-3} &= q_{k-1} \cdot r_{k-2} + r_{k-1}, & 0 < r_{k-1} < r_{k-2} \\
r_{k-2} &= q_k \cdot r_{k-1} + r_k, & 0 < r_k < r_{k-1} \\
r_{k-1} &= q_{k+1} \cdot r_k + 0; &
\end{aligned}$$

if $r_2 = 0$, then gcd$(a,b) = b$. Furthermore, there exist integers s and t such that

$$s \cdot a + t \cdot b = \gcd(a, b) \tag{1}$$

The extended Euclidean algorithm for integers. The integers s and t in (1) can be found by the *extended* Euclidean algorithm as follows. Let

$$s_0 = 1, \quad t_0 = 0, \quad r_0 = a$$

$$s_1 = 0, \quad t_1 = 1, \quad r_1 = b\,.$$

For $i \geq 2$, let q_i and r_i be defined by applying the division algorithm to r_{i-2} and r_{i-1}:

$$r_{i-2} = q_i \cdot r_{i-1} + r_i, \qquad \text{where } 0 \leq r_i < r_{i-1}. \tag{2}$$

Then compute

$$s_i = s_{i-2} - q_i \cdot s_{i-1}$$

$$t_i = t_{i-2} - q_i \cdot t_{i-1}\,.$$

The remainder sequence r_i is, of course, the same sequence defined in the (ordinary) Euclidean algorithm. The last nonzero remainder r_k here determines the index k specifying the multipliers $s = s_k$ and $t = t_k$ satisfying (1).

Example 1.

Application of the extended Euclidean algorithm to the integers 81 and 57 yields the following table:

i	s_i	t_i	r_i	q_i
0	1	0	81	–
1	0	1	57	–
2	1	–1	24	1
3	–2	3	9	2
4	5	–7	6	2
5	–7	10	3	1
6			0	2

Hence we see that $\gcd(81, 57) = 3$, and $-7 \cdot 81 + 10 \cdot 57 = 3$.

We now consider the case of polynomials in place of integers. As noted earlier, the division, Euclidean and extended Euclidean algorithms generalize directly. The restatement of the division algorithm follows.

The division algorithm for polynomials. Given a field F and polynomials $a(x), b(x) \in F[x]$, $b(x) \neq 0$, their exist unique polynomials $q(x), r(x) \in F[x]$ such that

$$a(x) = q(x) \cdot b(x) + r(x) \qquad \text{where } r(x) = 0 \text{ or } \deg r(x) < \deg b(x).$$

As in the integer case, the quotient polynomial $q(x)$ and remainder polynomial $r(x)$ can be found by standard "long division" of polynomials.

Euclidean algorithm and extended Euclidean algorithm for polynomials. The Euclidean and extended Euclidean algorithms generalize in the same way as the division algorithm, with the integer a replaced by the polynomial $a(x)$, etc., and with bounds on the remainders applying to the degree of the polynomial rather than the absolute value as in the integer case. For example, (2) generalizes to

$$r_{i-2}(x) = q_i(x) \cdot r_{i-1}(x) + r_i(x), \qquad \text{where } r_i(x) = 0 \text{ or } \deg r_i(x) < \deg r_{i-1}(x).$$

The multipliers s and t in (1), of course, also become polynomials $s(x)$ and $t(x)$. Regarding the computation of a gcd of two polynomials over finite fields, to ensure uniqueness of a gcd, by convention we take as the gcd the *monic* polynomial of largest degree which divides both polynomials. The extended Euclidean algorithm for polynomials is further discussed and utilized in §6.4.

Example 2.

Consider polynomials in $GF(5)[x]$. Applying the extended Euclidean algorithm to the polynomials

$$a(x) = x^8 + 4x^7 + 4x^6 + 4x^5 + 4x^4 + 2x^3 + x + 2$$

$$b(x) = x^6 + 3x^5 + 4x^4 + 2x^3 + 3x^2 + 2,$$

we proceed as follows. First apply the division algorithm to $a(x)$ and $b(x)$, yielding

$$a(x) = (x^2 + x + 2) \cdot b(x) + (2x^5 + x^4 + 2x^2 + 4x + 3),$$

i.e. yields $q_2(x)$ and $r_2(x)$ for the table below. This facilitates computation of $s_2(x) = 1$ and $t_2(x) = -x^2 - x - 2 \in GF(5)[x]$. Next, applying the division algorithm to $b(x)$ and $r_2(x)$

yields

$$b(x) = (3x)\cdot r_2(x) + (4x^4 + x^3 + x^2 + x + 2),$$

i.e. yields $q_3(x)$ and $r_3(x)$ for the table below. This allows computation of

$$s_3(x) = s_1(x) - q_3(x)\cdot s_2(x)$$

$$= 0 - 3x \cdot 1 = -3x$$

and

$$t_3(x) = t_1(x) - q_3(x)\cdot t_2(x)$$

$$= 1 - 3x \cdot (-x^2 - x - 2) = 3x^3 + 3x^2 + x + 1.$$

Continuing by applying the division algorithm to $r_2(x)$ and $r_3(x)$ yields

$$r_2(x) = (3x+2)\cdot r_3(x) + (2x^2 + x + 4),$$

i.e. yields $q_4(x)$ and $r_4(x)$. Now

$$s_4(x) = s_2(x) - q_4(x)\cdot s_3(x)$$

$$= 1 - (3x+2)\cdot(-3x) = 4x^2 + x + 1$$

and

$$t_4(x) = t_2(x) - q_4(x)\cdot t_3(x)$$

$$= (-x^2 - x - 2) - (3x+2)\cdot(3x^3 + 3x^2 + x + 1) = x^4 + 4x + 1.$$

Finally, applying the division algorithm to $r_3(x)$ and $r_4(x)$ gives

$$r_3(x) = (2x^2 + 2x + 3)\cdot r_4(x) + 0,$$

yielding a zero remainder to terminate the algorithm. The steps in the algorithm can be summarized by the following table.

i	$s_i(x)$	$t_i(x)$	$r_i(x)$	$q_i(x)$
0	1	0	$a(x)$	–
1	0	1	$b(x)$	–
2	1	$-x^2-x-2$	$2x^5+x^4+2x^2+4x+3$	x^2+x+2
3	$-3x$	$3x^3+3x^2+x+1$	$4x^4+x^3+x^2+x+2$	$3x$
4	$4x^2+x+1$	x^4+4x+1	$2x^2+x+4$	$3x+2$
5			0	$2x^2+2x+3$

From the second last line in the table, we conclude that a greatest common divisor of $a(x)$ and $b(x)$ is $2x^2+x+4$, and that

$$r_4(x) = s_4(x)\cdot a(x) + t_4(x)\cdot b(x).$$

The monic gcd of $a(x)$ and $b(x)$ is $\gcd(a(x), b(x)) = 3\cdot r_4(x) = x^2 + 3x + 2.$

Appendix C

THE CHINESE REMAINDER THEOREM

We state and prove here the Chinese remainder theorem, employed in §5.9. The reader is referred to [Berlekamp 68] for further details.

Chinese remainder theorem for polynomials. Given a field F and distinct irreducible polynomials $h_1(x), h_2(x), \ldots, h_r(x) \in F[x]$, where $h(x) = \prod_{i=1}^{r} h_i(x)$, and given polynomials $a_1(x), a_2(x), \ldots, a_r(x) \in F[x]$, the system of simultaneous congruences

$$f(x) \equiv a_i(x) \pmod{h_i(x)}$$

always has solutions for $f(x)$. Furthermore, $f(x)$ modulo $h(x)$ is unique.

Proof. Since the $h_i(x)$ are distinct and irreducible, $\gcd(\prod_{i \neq j} h_i(x), h_j(x)) = 1$. Hence by the Euclidean algorithm, there exist polynomials $s_j(x)$ and $t_j(x)$ such that

$$s_j(x) \cdot \prod_{i \neq j} h_i(x) \; + \; t_j(x) \cdot h_j(x) \; = \; 1 \, .$$

Let $H_j(x) = s_j(x) \cdot \prod_{i \neq j} h_i(x)$. Then

$$H_j(x) \equiv 1 \pmod{h_j(x)} \, . \tag{1}$$

Now define $f(x) = \sum_{i=1}^{r} a_i(x) \cdot H_i(x)$. Then since $h_i(x)$ divides $H_j(x)$ for $i \neq j$, it follows that

$$f(x) \equiv a_i(x) \cdot H_i(x) \pmod{h_i(x)}$$

and hence by (1),

$$f(x) \equiv a_i(x) \pmod{h_i(x)} \, .$$

To prove uniqueness, suppose $g(x)$ is another solution to the system. Then $g(x) \equiv a_i(x)$ $(\bmod\ h_i(x))$, implying

$$f(x) - g(x) \equiv 0 \pmod{h_i(x)}, \quad 1 \le i \le r$$

and hence

$$f(x) - g(x) \equiv 0 \ (\bmod \prod_{i=1}^{r} h_i(x)), \tag{2}$$

since the $h_i(x)$ are relatively prime. □ From (2) we have that $f(x) \equiv g(x) \pmod{h(x)}$.

Replacing the polynomials $h_i(x)$ in the above result by distinct prime numbers, and replacing the polynomials $a_i(x)$ by integers, yields the Chinese remainder theorem for integers; the same proof applies. Also, the polynomials $h_i(x)$ may be replaced by powers of irreducible polynomials, $h_i(x)^{e_i}$, as the proof relies only on the fact that the moduli are relatively prime.

Appendix D

FIELD REPRESENTATIONS AND ZECH'S LOG TABLES

The field $GF(2^3)$ can be generated by a root α of the irreducible polynomial $f(x) = 1 + x + x^3 \in Z_2[x]$. The vector representation of each field element is listed, with $(b_0b_1b_2)$ representing $b_0 + b_1\alpha + b_2\alpha^2$. Zech's logarithms $z(i)$ are also tabulated, where $1 + \alpha^i = \alpha^{z(i)}$.

$GF(2^3)$ generated by α where $f(x) = 1 + x + x^3$ and $f(\alpha) = 0$.

α^i	i	$z(i)$
000	∞	0
100	0	∞
010	1	3
001	2	6
110	3	1
011	4	5
111	5	4
101	6	2

The minimal polynomials of the field elements are given below. $m_i(x)$ denotes the minimal polynomial of α^i.

$$m_0(x) = 1 + x$$

$$m_1(x) = 1 + x + x^3$$

$$m_3(x) = 1 + x^2 + x^3$$

$GF(2^4)$ generated by α where $f(x) = 1 + x + x^4$ and $f(\alpha) = 0$.

α^i	i	$z(i)$
0000	∞	0
1000	0	∞
0100	1	4
0010	2	8
0001	3	14
1100	4	1
0110	5	10
0011	6	13
1101	7	9
1010	8	2
0101	9	7
1110	10	5
0111	11	12
1111	12	11
1011	13	6
1001	14	3

$m_0(x) = 1 + x$

$m_1(x) = 1 + x + x^4$

$m_3(x) = 1 + x + x^2 + x^3 + x^4$

$m_5(x) = 1 + x + x^2$

$m_7(x) = 1 + x^3 + x^4$

$GF(2^5)$ generated by α where $f(x) = 1 + x^2 + x^5$ and $f(\alpha) = 0$.

α^i	i	$z(i)$	α^i	i	$z(i)$
00000	∞	0	11111	15	24
10000	0	∞	11011	16	9
01000	1	18	11001	17	30
00100	2	5	11000	18	1
00010	3	29	01100	19	11
00001	4	10	00110	20	8
10100	5	2	00011	21	25
01010	6	27	10101	22	7
00101	7	22	11110	23	12
10110	8	20	01111	24	15
01011	9	16	10011	25	21
10001	10	4	11101	26	28
11100	11	19	11010	27	6
01110	12	23	01101	28	26
00111	13	14	10010	29	3
10111	14	13	01001	30	17

$m_0(x) = 1 + x$

$m_1(x) = 1 + x^2 + x^5$

$m_3(x) = 1 + x^2 + x^3 + x^4 + x^5$

$m_5(x) = 1 + x + x^2 + x^4 + x^5$

$m_7(x) = 1 + x + x^2 + x^3 + x^5$

$m_{11}(x) = 1 + x + x^3 + x^4 + x^5$

$m_{15}(x) = 1 + x^3 + x^5$

$GF(2^6)$ generated by α where $f(x) = 1 + x + x^6$ and $f(\alpha) = 0$.

α^i	i	$z(i)$	α^i	i	$z(i)$	α^i	i	$z(i)$
000000	∞	0	110111	21	42	010111	42	21
100000	0	∞	101011	22	50	111011	43	39
010000	1	6	100101	23	15	101101	44	37
001000	2	12	100010	24	4	100110	45	9
000100	3	32	010001	25	11	010011	46	30
000010	4	24	111000	26	7	111001	47	17
000001	5	62	011100	27	18	101100	48	8
110000	6	1	001110	28	41	010110	49	38
011000	7	26	000111	29	60	001011	50	22
001100	8	48	110011	30	46	110101	51	53
000110	9	45	101001	31	34	101010	52	14
000011	10	61	100100	32	3	010101	53	51
110001	11	25	010010	33	16	111010	54	36
101000	12	2	001001	34	31	011101	55	40
010100	13	35	110100	35	13	111110	56	19
001010	14	52	011010	36	54	011111	57	58
000101	15	23	001101	37	44	111111	58	57
110010	16	33	110110	38	49	101111	59	20
011001	17	47	011011	39	43	100111	60	29
111100	18	27	111101	40	55	100011	61	10
011110	19	56	101110	41	28	100001	62	5
001111	20	59						

$m_1(x) = 1 + x + x^6$

$m_3(x) = 1 + x + x^2 + x^4 + x^6$

$m_5(x) = 1 + x + x^2 + x^5 + x^6$

$m_7(x) = 1 + x^3 + x^6$

$m_9(x) = 1 + x^2 + x^3$

$m_{11}(x) = 1 + x^2 + x^3 + x^5 + x^6$

$m_{13}(x) = 1 + x + x^3 + x^4 + x^6$

$m_{15}(x) = 1 + x^2 + x^4 + x^5 + x^6$

$m_{21}(x) = 1 + x + x^2$

$m_{23}(x) = 1 + x + x^4 + x^5 + x^6$

$m_{27}(x) = 1 + x + x^3$

$m_{31}(x) = 1 + x^5 + x^6$

$GF(3^3)$ generated by α where $f(x) = 1 + 2x^2 + x^3$ and $f(\alpha) = 0$.

α^i	i	$z(i)$	α^i	i	$z(i)$
000	∞	0			
100	0	13	200	13	∞
010	1	18	020	14	24
001	2	7	002	15	16
201	3	2	102	16	20
221	4	12	112	17	23
220	5	14	110	18	11
022	6	21	011	19	22
101	7	3	202	20	15
211	8	19	122	21	9
222	9	6	111	22	8
121	10	4	212	23	25
210	11	1	120	24	5
021	12	10	012	25	17

$m_0(x) = 2 + x$

$m_1(x) = 1 + 2x^2 + x^3$

$m_2(x) = 2 + 2x + 2x^2 + x^3$

$m_4(x) = 2 + 2x + x^3$

$m_5(x) = 1 + 2x + x^2 + x^3$

$m_7(x) = 1 + x + 2x^2 + x^3$

$m_8(x) = 2 + x + x^2 + x^3$

$m_{13}(x) = 1 + x$

$m_{14}(x) = 2 + x^2 + x^3$

$m_{17}(x) = 1 + 2x + x^3$

REFERENCES

[Berlekamp 68] E.R. Berlekamp, *Algebraic Coding Theory*, McGraw-Hill, 1968.

[Blahut 83] R.E. Blahut, *Theory and Practice of Error Control Codes*, Addison-Wesley, 1983.

[Blake 76] I.F. Blake and R.C. Mullin, *An Introduction to Algebraic and Combinatorial Coding Theory*, Academic Press, 1976.

[Gilbert 76] W.J. Gilbert, *Modern Algebra with Applications*, Wiley, 1976.

[Hill 86] R. Hill, *A First Course in Coding Theory*, Oxford University Press, 1986.

[Hoeve 82] H. Hoeve, J. Timmermans and L.B. Vries, "Error correction and concealment in the Compact Disc system," *Philips Technical Review* 40, No. 6, 166-172, 1982.

[Lidl 84] R. Lidl and H. Niederreiter, *Finite Fields*, Cambridge University Press, 1984.

[Lin 83] S. Lin and D.J. Costello, Jr., *Error Control Coding: Fundamentals and Applications*, Prentice-Hall, 1983.

[Lint 82] J.H. van Lint, *Introduction to Coding Theory*, Springer-Verlag, 1982.

[Lint 86] J.H. van Lint and R.M. Wilson, "On the minimum distance of cyclic codes", *IEEE Transactions on Information Theory* Vol. IT-32, No. 1 (1986), 23-40.

[MacWilliams 77] F.J. MacWilliams and N.J.A. Sloane, *The Theory of Error-Correcting Codes*, North-Holland, 1977.

[McEliece 77] R.J. McEliece, *The Theory of Information and Coding*, Addison-Wesley, 1977.

[McEliece 87] R.J. McEliece, *Finite Fields for Computer Scientists and Engineers*, Kluwer Academic Publishers, 1987.

[Peterson 72] W.W. Peterson and E.J. Weldon, Jr., *Error-Correcting Codes*, MIT Press, 1972.

[Pless 82] V. Pless, *Introduction to the Theory of Error-correcting Codes*, Wiley, 1982.

[Pohlmann 87] K.C. Pohlmann, *Principles of Digitial Audio*, Howard W. Sams & Co. (A Division of Macmillan, Inc.), 1987.

[Vries 82] L.B. Vries and K. Odaka, "CIRC - The Error Correcting Code for the Compact Disc Digital Audio System," in *Digital Audio* (Collected Papers from the AES Premier Conference, Rye, New York, 1982 June 3-6), Audio Engineering Society Inc., 1983.

Subject Index